ERGEBNISSE DER
ANGEWANDTEN MATHEMATIK

UNTER MITWIRKUNG DER SCHRIFTLEITUNG DES
„ZENTRALBLATT FÜR MATHEMATIK"
HERAUSGEGEBEN VON
F. L. BAUER · L. COLLATZ · F. LÖSCH · C. TRUESDELL

———————————— 7 ————————————

MATHEMATICAL THEORY
OF ELASTIC EQUILIBRIUM
(RECENT RESULTS)

BY

GIUSEPPE GRIOLI

SPRINGER-VERLAG
BERLIN · GÖTTINGEN · HEIDELBERG
1962

ISBN-13: 978-3-540-02804-8   e-ISBN-13: 978-3-642-87432-1
DOI: 10.1007/978-3-642-87432-1

Preface

It is not my intention to present a treatise of elasticity in the following pages. The size of the volume would not permit it, and, on the other hand, there are already excellent treatises. Instead, my aim is to develop some subjects not considered in the best known treatises of elasticity but nevertheless basic, either from the physical or the analytical point of view, if one is to establish a complete theory of elasticity.

The material presented here is taken from original papers, generally very recent, and concerning, often, open questions still being studied by mathematicians. Most of the problems are from the theory of finite deformations [non-linear theory], but a part of this book concerns the theory of small deformations [linear theory], partly for its interest in many practical questions and partly because the analytical study of the theory of finite strain may be based on the infinitesimal one.

In the linear theory I wish to consider [Chapter VI] rather broadly the problem of establishing some inequalities for the stresses and displacements [maximum stress, etc.]. In fact, the aim is to find maximum or minimum values of these quantities, which are often sufficient for practical applications, while the complete knowledge of certain functions which may be obtained only by integration of the analytical problem, is superfluous. General integration methods may be established in the linear case by introducing polynomial approximations for the displacement components or, directly, for the stress components. It is interesting to observe that such methods, particularly adapted to numerical solution by means of modern electronic calculating machines, are based on the application of well known theorems of physics and mathematics, such as BETTI's and MENABREA's theorems.

Further, starting from such integration methods for the linear problem it is even possible to establish an integration procedure for the static problem of finite deformations [Chapter VII]. This problem presents great analytical difficulties and has not yet been settled completely in regard to existence, uniqueness, series expansion, etc. Only when the external forces are all known [constraints absent] may the solution now be considered completed; for this case interesting theorems based on modern functional analysis have been established [Chapter V].

The complexity of the non-linear problem is such that the possibility of transforming it into a succession of analytical linear problems [Chapter IV] makes the study of classical linear elasticity fundamental even for the non-linear case.

Further, such a reduction of the basic problem of finite deformations permits the study of several subtle questions; for example, the possibility of removing the classical indeterminacy of the rigid displacement, present in the linear theory when constraints are absent or allow such a displacement. Further, one sees that the meaning of solutions of the linear theory—at least of those belonging to a certain analytical class—in a certain exceptional case may be doubtful [Chapter IV].

A brief chapter is dedicated to the linear problem of plane elasticity [Chapter VIII]. Substantially, it contains some considerations of the AIRY stress function and its analytical structure even when dislocations are present, with reference to the case of plane deformation and to the very restrictive case of plane stress.

A natural idea, often accepted, regarding the analytical structure of the thermodynamic potential is that of expanding it in power series of the parameters on which it depends, taking only a few terms. Such an approximate expression, depending on a certain number of coefficients which are to be determined, is substituted for the correct one. I do not take up this subject. Instead, I have considered some types of thermo-dynamic potential [Chapter III] which have non-approximate character but are based on hypotheses of analytical or physical simplicity, parti-cularly the *elasticity of second grade*, which, based on the hypothesis that the Eulerian stress components are quadratic functions of the deforma-tion components, presents some interesting features.

Special questions are considered in the last two chapters.

The first concerns *hypo-elasticity* [Chapter IX]. This recent continuum theory is based on renunciation of the idea of a reference natural state, assuming instead that stress and strain are connected through con-stitutive equations which relate the rate of stress to the rate of strain. For this reason some difficulties may arise in the static case, but interest-ing solutions have already been obtained.

The theory has broader possibilities than the classical non-linear elasticity, since some plastic phenomena, perhaps, may belong to the field of hypo-elasticity and, further, an arbitrary stress may be present in the initial state. Doubtless, not only the Eulerian point of view but also the Lagrangean one may be convenient in hypo-elasticity.

The last chapter concerns a much neglected field: *asymmetric elasti-city*.

The hypothesis of asymmetric stress components has always been considered interesting only in special cases; for example, when body moments are present [electromagnetic phenomena]. However, classical elasticity theory often gives solutions with singularities due exclusively to the fact that the assigned external surface forces are such as to imply an asymmetry of the stress components, at least in certain parts of the body.

Such singularities are not physically plausible and are contrary to the experimental results. On the other hand, these singularities disappear in a theory which admits the possibility that the stress components may be asymmetric even if body moments are absent.

Internal constraints are excluded in all questions considered. In particular, for brevity, incompressible bodies, for which a large literature exists, are not considered.

The first two chapters are of introductory character, to help in understanding the questions considered in the others.

The bibliography at the end of the book lists only those papers which have direct bearing on the subjects discussed or are basic to their study; it is not intended to be exhaustive.

Table of Contents

Chapter I

Kinematic Introduction

§ 1. Preliminaries

1. By the term three-dimensional continuous body, S, I refer to a natural body which can be represented mathematically as follows: a) S fills a region C of three-dimensional space, b) S has mass m, defined by

$$m = \int_C k\, dC \tag{1.1}$$

where k is the density.

By C I mean not only the region occupied by S but also the configuration of S, that is, not only its shape but also that of the various parts into which one can imagine it divided.

Out of the possible configurations of the body select two, denoted by C^* and C. Refer the body to a rectangular Cartesian frame T with origin O and unit vectors c_1, c_2, c_3 and denote by y_1, y_2, y_3 the coordinates of an arbitrary point P^* of C^*, and by x_1, x_2, x_3 those of an arbitrary point P of C. Between the two possible configurations C^*, C there is a one-to-one mapping which sets P into correspondence with P^*. By this I mean that the x_r are functions of the y_r,

$$x_r = x_r\,(y_1, y_2, y_3)\,, \tag{1.2}$$

which satisfy within and on the boundary of every measurable set in C^* the following conditions[1]:

a) They are bounded, continuous and single-valued.

b) They have derivatives up to and including the second order except on a set of measure zero.

c) They are invertible.

d) Their Jacobian determinant D is positive:

$$D = \|\,x_{r,s}\,\| > 0\,. \tag{1.3}$$

---

[1] Exceptional cases in which these conditions are not satisfied, as in VOLTERRA's dislocations, are generally excluded.

Indices following a comma denote derivatives with respect to $y_r \left[x_{r,s} = \dfrac{\partial x_r}{\partial y_s} \right]$, and every index runs from 1 to 3 unless the contrary is stated. Naturally, in the second members of (1.2) some other parameter may occur, such as the time or any other parameter serving to individualize one configuration from an assigned succession. When not wishing to emphasize such dependence, as at present, I shall use the notation (1.2) without explicit appearance of any parameter. The correspondence (1.2) establishes others between elements of C^* and elements of C such as lengths, areas, etc. I shall use the same letters, with or without asterisks, to indicate elements of C^* or C, respectively, putting asterisks above or below, indifferently.

Generally I shall omit summation symbols, intending, as usual, that sums shall always be understood when there are two like indices, one of which is a subscript and the other a superscript. Many of the symbols refer to tensors. Indices will be written above or below as preferred. By this I do not mean to indicate covariant or contravariant character, since the coordinates are rectangular Cartesian and the space is Euclidean.

§ 2. Fundamental quantities describing the deformation

1. To describe the deformation of S in the transformation from C^* to C the tensor

$$b_{rs} = x_{i,r}\, x^i_{,s} \tag{1.4}$$

is fundamental. The squared element of arc in C^*,

$$ds^2_* = dy_i\, dy^i\,, \tag{1.5}$$

corresponds to the squared element of arc in C,

$$ds^2 = b_{rs}\, dy^r\, dy^s\,. \tag{1.6}$$

Also fundamental is the tensor ε_{rs}, related to b_{rs} by the equation

$$b_{rs} = \delta_{rs} + 2\,\varepsilon_{rs} \tag{1.7}$$

where δ_{rs} denotes the Kronecker delta.

The extension of linear elements in the direction of the unit vector c_i is

$$\delta_i = \sqrt{b_{ii}} - 1 = \sqrt{1 + 2\varepsilon_{ii}} - 1 > -1\,. \tag{1.8}$$

If ϑ_{ih} is the angle between two linear elements of C corresponding to two linear elements of C^* which are parallel to the unit vectors c_i, c_h, it follows, as is known, that

$$\sin \tau_{ih} = \cos \vartheta_{ih} = \frac{b_{ih}}{\sqrt{b_{ii}\, b_{hh}}} = \frac{\varepsilon_{ih}}{\sqrt{(1 + 2\varepsilon_{ii})\,(1 + 2\varepsilon_{hh})}} \qquad (i \neq h) \tag{1.9}$$

where $\tau_{ih} = \dfrac{\pi}{2} - \vartheta_{ih}$.

Let u_i be the components of the displacement P^*P, so that

$$x_i = y_i + u_i , \qquad (1.10)$$

and

$$\varepsilon_{rs} = \frac{1}{2} \left(x_{i,r} \, x^{i,s} - \delta_{rs} \right) = \frac{1}{2} \left(u_{r,s} + u_{s,r} + u_{i,r} \, u^i_{,s} \right) . \qquad (1.11)$$

It will often be useful to use for ε_{rs} a single-index notation in which the *components of strain* ε_r $(r = 1, 2, ..., 6)$ are defined by

$$\varepsilon_r = \varepsilon_{rr} , \qquad \varepsilon_{r+3} = 2 \, \varepsilon_{r+1 \, r+2} ; \qquad (1.12)$$

of course if in $\varepsilon_{r+1 \, r+2}$ an index exceeds 3, it should be diminished by 3.

The condition $ds = ds^*$ for each linear element of C^* is necessary and sufficient that the transformation be rigid. Consequently, an alternative necessary and sufficient condition is $b_{rs} = \delta_{rs}$, or, equivalently, $\varepsilon_{rs} = 0$.

The quadratic form (1.6) must be reducible to a sum of squares. For this it is necessary and sufficient that the Riemann tensor formed from b_{rs} be zero. Its distinct components are six: $R_{1212}, R_{2323}, R_{3131}, R_{1213}, R_{1223}, R_{2331}$. Setting them equal to zero yields[1] six conditions of integrability for b_{rs} or for ε_{rs}. The linear part of these equations gives the well known Saint-Venant conditions of compatibility for infinitesimal strain.

2. Set

$$e_{rs} = \frac{1}{2} \left(u_{r,s} + u_{s,r} \right) , \qquad \omega_r = \frac{1}{2} \left(u_{r+2,r+1} - u_{r+1,r+2} \right) . \qquad (1.13)$$

The e_{rs} are the usual expressions substitued for ε_{rs} in the linear theory, while the ω_r are the components of local rotation in this theory. In finite strain, e_{rr} may be interpreted as the extension of an element of arc from P^* parallel to the axis of y_r, provided that before the comparison the deformed element is projected onto the same axis.

Also ω_r has a meaning in the case of large strain: ω_r is proportional to the integral of $\operatorname{tg} \Psi_r$ where Ψ_r is the angle of rotation around the axis of the unit vector c_r of elements of arc from P^* perpendicular to c_r [NOVO-ZHILOV, 1, p. 31]. Indeed,

$$\frac{1}{2\pi} \int_0^{2\pi} \operatorname{tg} \Psi_r \, d\vartheta' = \frac{\omega_r}{\sqrt{(1 + 2 e_{r+1 r+1})(1 + e_{r+2 r+2}) - e^2_{r+1 r+2}}} , \qquad (1.14)$$

where ϑ' is the angle made by the linear element with the unit vector c_{r+1} .

---

[1] The conditions of integrability for the tensor of deformation are given explicitly in [PLATRIER, C.].

It is interesting to observe that ε_r may be expressed in terms of e_{rs} and ω_r. By (1.11), (1.12), (1.13) it follows, in fact, that

$$\varepsilon_r = e_{rr} + \frac{1}{2} \left[e_{rr}^2 + (e_{rr+1} + \omega_{r+2})^2 + (e_{rr+2} - \omega_{r+1})^2 \right],$$

$$\varepsilon_{r+3} = 2e_{r+1\,r+2} + e_{r+1\,r+1} \left(e_{r+1\,r+2} - \omega_{r+3} \right) + e_{r+2\,r+2} \left(e_{r+1\,r+2} + \omega_{r+3} \right) +$$

$$+ \left(e_{r+1\,r+3} - \omega_{r+2} \right) \left(e_{r+2\,r+3} + \omega_{r+1} \right). \tag{1.15}$$

From (1.15) it is easy to see conditions when it is justified to identify ε_r, ε_{r+3} with the linear expression e_{rs}, as is usually done in the classical theory of elasticity. In [NOVOZHILOV, 1] there is a detailed examination of the question. Here I recall only that one can simplify (1.15) in two cases: a) when the displacements are small in comparison with the size of the body and the rotations are small in comparison with 1; b) when the extensions and shears are small in comparison with 1. Case a) is the one substantially accepted in the classical theory of bodies subject to small deformations. It is the more restrictive, since it implies case b) but not vice versa. In case a) ε_{rs} are identified with e_{rs}; in case b), however, such identification is not generally justified, as, for example, in theories of plates, beams, shells, etc.

3. Let dP^* and δP^* be two non-parallel infinitesimal vectors of C^* emanating from P^*; let their coordinates be dy_i, δy_i, and let dP, δP be the corresponding elements of C. Let dA^*, dA be the areas of the corresponding parallelograms, and let n^*, n be unit vectors perpendicular to planes of dA^*, dA. I presume that the vectors dP^*, δP^*, n^* form a right-handed set. Then

$$n_h^* dA^* = \varepsilon_{hik}\, dy^i \delta y^k, \qquad n_h dA = \varepsilon_{hik}\, x^i_{,s}\, x^k_{,t}\, dy^s \delta y^t, \tag{1.16}$$

where ε_{ihk} is Ricci's tensor.

Let C_{rs} be the cofactor of $x_{r,s}$ in the matrix $|x_{r,s}|$. The equality

$$C_{rs} = \varepsilon_s^{\cdot ih}\, x_{r+1,\,i}\, x_{r+2,\,h} \tag{1.17}$$

shows that (1.16, 2) may be written in the form

$$n_r dA = C_{rs}\, n_*^s dA^*. \tag{1.18}$$

From (1.18) it follows that the superficial dilatation δ_{An} for elements of area perpendicular to n^* is

$$\delta_{An} = \sqrt{C_{rs}\, C_{\cdot l}^r\, n_*^s\, n_*^l} - 1. \tag{1.19}$$

It may be demonstrated that (1.19) is equivalent to

$$\delta_{An} = \sqrt{B_{rs}\, n_*^r\, n_*^s} - 1 \tag{1.20}$$

where B_{rs} denotes the cofactor of b_{rs} in the matrix $|b_{rs}|$. Thus it is clear that b_{rs} [or ε_{rs}] characterizes not only extensions and shears but also superficial dilatations. The same thing happens for cubical dilatation. In fact, the well known expression for the cubical dilatation,

$$\delta_c = D - 1 , \tag{1.21}$$

is equivalent to

$$\delta_c = \sqrt{B} - 1 , \tag{1.22}$$

where B denotes the determinant, necessarily positive, of b_{rs}.

4. The symmetric tensor b_{rs} satisfies the conditions $b_{rr} > 0$ and $B > 0$. These are sufficient to assure that at each $P*$ there is at least one set T_B of principal directions for the tensor b_{rs}. This means that the vector $b_{rs} v^s$ is parallel to the vector v, if this latter is parallel to one of the vectors of T_B. From this follows the existence of a tensor d_{rs} satisfying the equality

$$d_{rt} d^t_{.s} = b_{rs} . \tag{1.23}$$

The tensor d_{rs} has the same principal directions as b_{rs}, and its proper numbers are the square roots of the proper numbers of the tensor b_{rs}. Therefore, in a principal frame at the $P*$ in question

$$d_{rr} = \sqrt{b_{rr}} > 0 , \qquad d_{rs} = d_{sr} = 0 \qquad (r \neq s) . \tag{1.24}$$

In an arbitrary reference frame T, (1.24) are not valid, but from (1.23) follows

$$D' d_{sq} = D^t_{.s} b_{tq} , \tag{1.25}$$

where D' is the determinant[1] $\|d_{rs}\|$ and D_{rs} is the cofactor of d_{rs}. Let

$$\varrho_{rp} = \frac{1}{D'} x_{r,q} D_p^{.q} . \tag{1.26}$$

The tensor ϱ_{rp} satisfies the equality

$$x_{r,s} = \varrho_{rt} d^t_{.s} . \tag{1.27}$$

The tensors d_{rs}, D_{rs} are symmetric, and from (1.25), (1.26) it follows that

$$D'^2 \varrho^r_{.s} \varrho_{rp} = x^r_{.t} x_{r,q} D^t_{.s} D^q_{.p} = b_{tq} D^t_{.s} D^q_{.p} = D' d_{sq} D^q_{.p} = D'^2 \delta_{sp} . \tag{1.28}$$

That is,

$$\varrho^r_{.s} \varrho_{rp} = \delta_{sp} . \tag{1.29}$$

---

[1] It is easy to show that $D' = D$.

In other words ϱ_{rs} represents a rigid rotation, and (1.27) shows an important decomposition of the displacement $C^* \to C$: except in a translation, the displacement of every infinitesimal element of C^* is the superposition of a pure deformation, characterized by the tensor d_{rs}, and of a rigid rotation characterized by ϱ_{rs}. That is, an infinitesimal vector at P^*, having coordinates dy_i, is changed into a vector at P whose coordinates are

$$dx^i = \varrho^{it} d_{ts} dy^s .\tag{1.30}$$

It is easy to solve (1.23) if $b_{13} = b_{23} = 0$ [SIGNORINI, 3, p. 46]. In particular, this hypothesis is verified in the cases of pure bending and of pure shear. Let

$$\Delta_3 = b_{11} b_{12} - b_{12}^2 , \qquad R = b_{11} + b_{22} + \sqrt{\Delta_3} .\tag{1.31}$$

From (1.23) follows

$$d_{ii} = \frac{b_{ii} + \sqrt{\Delta_3}}{\sqrt{R}} \qquad (i = 1, 2) ,$$

$$d_{33} = \sqrt{b_{33}} , \quad d_{12} = \frac{b_{12}}{\sqrt{R}} , \qquad d_{13} = d_{23} = 0 .\tag{1.32}$$

Then (1.26) give the tensor ϱ_{rs}.

§ 3. A minimum theorem in the kinematics of large deformations

If displacement $C^* \to C$ is *homogeneous*, that is, if $x_{r,s}$ are independent of y_t, then d_{rs}, ϱ_{rs} are independent of y_t [SIGNORINI, 3, p. 54]. The displacement in a sufficiently small portion of C^* around P^* is identified with the *homogeneous tangent displacement* in P^*, that is with the homogeneous displacement characterized by the translation P^*P and by the values of d_{rs}, ϱ_{rs} at P^*. Let c_* be a very small spherical portion of C^* around P^*, let Q^* be a point of c_* and let Q', Q'' be the corresponding points of Q^* in two different displacements S', S'' starting from C^*. I call

$$d_{p*} = \int\limits_{c_*} |Q' Q''|^2 dC^* \tag{1.33}$$

the *local deviation* of the two displacements S', S'' with reference to P^*. The following theorem holds: *the rigid displacement minimizing the local deviation at P^* from the real displacement is just the rigid part of the homogeneous tangent displacement at P^** [GRIOLI, 1]. In other words, the rigid displacement which at P^* best approximates the real one is the product of the translation P^*P and the rotation characterized by the values of ϱ_{rs} at P^*. To demonstrate the theorem one identifies S' with the homogeneous tangent displacement at P^*, which approximates the real displacement in c_*, and S'' with an arbitrary rigid displacement, the product of a translation of the vector t'' by a rotation expressed by the tensor ϱ''_{rs}.

Putting at P^* the origin of the reference frame T, we have

$$P^*Q' = P^*P + PQ' = P^*P + \varrho^{rt} d_{ts} y^s c_r ,$$
$$P^*Q'' = t'' + \varrho''^{rs} y_s c_r , \tag{1.34}$$

and therefore

$$Q'Q'' = t'' - P^*P + (\varrho''^{rs} - \varrho^{rt} d_t^{\cdot s}) y_s c_r . \tag{1.35}$$

Let τ be the length of the vector $t'' - P^*P$, and let τ_r be its coordinates. From (1.35) follows

$$|Q'Q''|^2 = \tau^2 + y_l y^l + d_q^{\cdot s} d^{ql} y_s y_l + 2\tau (\varrho''^{rs} - \varrho^{rt} d_t^{\cdot s}) y_s - $$
$$- 2 \varrho''^{rs} \varrho_r^{\cdot t} d_t^{\cdot l} y_s y_l . \tag{1.36}$$

Therefore, according to (1.23),

$$d_{p^*} = \frac{4}{3} \pi a^3 \tau^2 + \frac{4}{15} \pi a^5 [b_r^{\cdot r} + 3 - 2 \varrho''^{rs} \varrho_r^{\cdot t} d_{ts}] , \tag{1.37}$$

where a denotes the radius of c^*. If the reference frame, T, is a principal frame for b_{rs}, then $d_{ts} = 0$ $(t \neq s)$, and (1.37) becomes

$$d_{p^*} = \frac{4}{3} \pi a^3 \tau^2 + \frac{4}{15} \pi a^5 [b_r^{\cdot r} + 3 - 2 \sum_s \varrho''^{rs} \varrho_r^{\cdot s} \sqrt{b_{ss}}] . \tag{1.38}$$

The tensor $\varrho''^{rs} \varrho_r^{\cdot t}$ represents a rigid rotation. Therefore

$$\varrho''^{rs} \varrho_r^{\cdot s} \leqslant 1 . \tag{1.39}$$

Thus, from (1.38) follows

$$d_{p^*} \geqslant \frac{4}{15} \pi a^5 [b_r^{\cdot r} + 3 - 2 \sum_s \sqrt{b_{ss}}] = \frac{4}{15} a^5 \pi \sum_s \Delta_s^2 , \tag{1.40}$$

where the Δ_s are the principal extensions at P^*. In (1.40) the sign of equality is valid if and only if $\tau = 0$ and if (1.39) is an equality. This means that d_{p^*} is a minimum if and only if

$$t'' = P^*P , \qquad \varrho''_{rs} = \varrho_{rs} . \tag{1.41}$$

Therefore S'' coincides with the homogeneous tangent displacement in P^*, as was to be demonstrated.

Chapter II

Basic Equations of the Statics of Continuous Media

§ 1. Eulerian expressions of fundamental equations

Let π be the plane perpendicular to the unit vector $\boldsymbol{u}$ from the interior point P of C, and let $d\sigma$ be an infinitely small region of π including P. π divides the body into two parts, and I shall call positive that part which contains $\boldsymbol{u}$, and the other one negative. In the mathematical theory of continuous media it is fundamental to consider the contact forces which the particles of the body near π and belonging to the negative part exert across $d\sigma$ on the particles belonging to the other part.

It is supposed that those forces are equivalent to their resultant $\boldsymbol{R}_u$ applied at P. This resultant is proportional to $d\sigma$. This means that the limit of $\dfrac{\boldsymbol{R}_u}{d\sigma}$ generally is neither zero nor infinite as $d\sigma \to 0$, while the limit of the quotient by $d\sigma$ of the moment about P of those contact forces is zero. In my opinion, the case when this latter limit is not zero is interesting, and I will consider it in Chapter X.

Let

$$\boldsymbol{R}_u = \boldsymbol{\Phi}_u d\sigma . \tag{2.1}$$

Let ΔC be any portion of C entirely in the interior of C, let σ be its boundary, let $\boldsymbol{n}$ be the unit vector parallel to the interior normal to σ and let n_s be the coordinates of $\boldsymbol{n}$. The basic equations of statics are

$$\int_{\Delta C} k\boldsymbol{F} dC + \int_{\sigma} \boldsymbol{\Phi}^s n_s d\sigma = 0 , \tag{2.2}$$

$$\int_{\Delta C} OP \times k\boldsymbol{F} dC + \int_{\sigma} OP \times \boldsymbol{\Phi}^s n_s d\sigma = 0 , \tag{2.3}$$

where[1] $k\boldsymbol{F}$ denotes the body force per unit mass, and where $\boldsymbol{\Phi}^s$ is the vector $\boldsymbol{\Phi}_u$ if $\boldsymbol{u}$ is parallel to c_s. From (2.2) Cauchy's equations follow:

$$X'^s_{rs} = kF_r \qquad \text{(in } C\text{)} , \tag{2.4}$$

$$X_{rs} N^s = f_r \qquad \text{(on } \Sigma\text{)} , \tag{2.5}$$

where

$$X_{rs} = c_r \cdot \boldsymbol{\Phi}_s . \tag{2.6}$$

In (2.5) N^s are the direction cosines of the interior normal $\boldsymbol{N}$ to the boundary Σ of C. From (2.3) follow the conditions of symmetry:

$$X_{rs} - X_{sr} = 0 . \tag{2.7}$$

---

[1] As usual, the sign $\times$ denotes the vector product of two vectors, the sign $\cdot$ the scalar product.

It is proper to remark that (2.4) or (2.7) may fail to hold on a set of measure zero in C.

The work done by the contact forces for any infinitesimal displacement starting from C is expressed by

$$\delta L^{(i)} = \int_C \delta l^{(i)} \, dC , \qquad (2.8)$$

with

$$\delta l^{(i)} = \frac{1}{2} [(\delta u_r)_{,s} + (\delta u_s)_{,r}] X^{rs} , \qquad (2.9)$$

where δu_r are the components of the infinitesimal displacement.

§ 2. Lagrangean expressions of fundamental equations

1. Let

$$X_{rs} = \frac{1}{D} Y_{lm} x_r^{'l} x_s^{'m} ; \qquad (2.10)$$

equivalently,

$$Y_{rs} = Y_{sr} = \frac{1}{D} X_{lm} C_{.r}^l C_{.s}^m . \qquad (2.11)$$

It is well known that (2.4), (2.5) are equivalent to Lagrangean equations[1]

$$(Y^{ts} x_{r,i})_{,s} = k^* F_r \qquad \text{(in } C^*\text{)} , \qquad (2.12)$$

$$Y^{ts} x_{r,i} N_s^* = f_r^* \qquad \text{(on } \Sigma^*\text{)} , \qquad (2.13)$$

where the vector F_r is to be regarded as a function of y_t, while the vector f^* is expressed by the equality

$$f^* = f (1 + \delta_\Sigma) . \qquad (2.14)$$

In (2.14) δ_Σ denotes the superficial dilatation at a point of Σ, with reference to the tangent plane. The Y_{rs}, introduced by PIOLA, are the Lagrangean components of stress. From the equality

$$X^{rl} n_l d\sigma = Y^{ts} x_{,t}^r n_s^* d\sigma^* \qquad (2.15)$$

it is easy to recognize that the transformation $C^* \to C$ changes the stress whose components are $Y^{ts} n_s^* d\sigma^*$ and which corresponds to $n^* d\sigma^*$ into the stress whose components are $X^{rl} n_l d\sigma$ and which corresponds to $n d\sigma$. This justifies the name of Lagrangean components of stress for Y_{rs}. Multiplying (2.12), (2.13) by $x^{i,r}$, by easy calculations one finds that the

---

[1] Equations (2.12), (2.13) have been known since the last century. For brevity, I omit listing other Lagrangean forms of the basic equations [BOUSSINESQ, KIRCHHOFF]. For the various forms of the equations of statics in curvilinear coordinates see [TONOLO, 1, 2].

resulting equations may be written in the form[1]

$$Y_{st}^{\cdot s}\left(\delta^{rt}+2\,\varepsilon^{rt}\right)+\varepsilon^{rt,s}\,Y_{st}+\left(\varepsilon_s^{r,\,r+2}-\varepsilon_s^{r+2,r}\right)Y^{r+2s}+$$
$$+\left(\varepsilon_s^{r,\,r+1}-\varepsilon_s^{r+1,r}\right)Y^{r+1s}=k^*\,x^{t,r}\,F_t\,,\qquad(2.16)$$
$$\left(\delta^{rt}+2\,\dot\varepsilon^{rt}\right)Y_{ts}\,N_*^s=x^{t,r}\,f_t^*\;.$$

Let us suppose known a relation between Y_{rs} and the ε_{pq} alone, expressed by the equalities

$$Y_{rs}=\varphi_{rs}\left(\varepsilon_{pq}\right),\qquad \varepsilon_{rs}=\varphi_{rs}'\left(Y_{pq}\right),\qquad(2.17)$$

as is the case, for example, for elastic bodies under adiabatic or isothermal transformations.

From (2.17) follows that it is possible to express the left members of (2.16) in terms of the Y_{rs} alone. Not so for the right members of (2.16). Such an expression is possible for (2.16, 1) only if body forces are zero; for (2.16,2), only if surface forces are absent.

Often it will be convenient to use for Y_{rs} symbols with only one index, putting

$$Y_r=Y_{rr}\,,\qquad Y_{r+3}=Y_{r+1\,r+2}\,,\qquad(2.18)$$

and likewise for X_{rs}.

2. The Lagrangean expression of the work of the internal forces is

$$\delta L^{(i)}=\int_{C^*}\delta^*l^{(i)}\,dC^*\qquad(2.19)$$

where

$$\delta^*l^{(i)}=\frac{1}{2}\,Y_{rs}\,\delta b^{rs}=Y_{rs}\,\delta\varepsilon^{rs}=\sum_{j=1}^{6}Y_j\,\delta\varepsilon^j\;.\qquad(2.20)$$

In (2.20), δb^{rs}, $\delta\varepsilon^{rs}$, $\delta\varepsilon^j$ denote respectively the variations of b_{rs}, ε_{rs}, ε^j [see (1.4), (1.7), (1.12)] corresponding to an infinitesimal displacement from C.

§ 3. Thermodynamic potential

1. Let U be the internal energy, E the entropy, T the absolute temperature and e the mechanical equivalent of heat. Let

$$J=U-eTE\qquad(2.21)$$

be the *free energy of Helmholtz*.

If the body is a system *with reversible transformations*, for every infinitesimal transformation the equality

$$\frac{\delta Q}{T}=\delta E\qquad(2.22)$$

---

[1] Equations (2.16) correspond to [SIGNORINI, 3, p. 112].

is valid, where δQ denotes the heat absorbed by the body. By well known considerations based on the first and second laws of thermodynamics it follows that

$$\delta * l^{(i)} + e E \delta T = - \delta J . \tag{2.23}$$

If there is no inner constraint, (2.20) and (2.23) show that J depends on the actual state only through ε_j and T, and it follows that[1]

$$Y_j = - \frac{\partial J}{\partial \varepsilon_j}, \quad (j = 1, 2, \ldots, 6); \quad eE = - \frac{\partial J}{\partial T}. \tag{2.24}$$

For elastic bodies there exists at least one configuration of *free equilibrium* $\bar{C}$, in which equilibrium holds when all external forces vanish, such that the work of internal forces from $\bar{C}$ is always negative for any isothermal and non-rigid displacement, if the temperature lies in a certain interval. Then assuming $C* \equiv \bar{C}$ and supposing $J = 0$ in $C*$, we infer the inequality

$$\int_{C*} J (\varepsilon_{pq}, T*) \, dC* > 0 , \tag{2.25}$$

for any displacement from $C*$ which is not a rigid one. Independently of thermodynamic considerations, one may observe that in order to show the differential (2.20) to be exact it is sufficient to make the hypothesis that $\int \delta * l^{(i)}$, as expressed by (2.20), is not positive for any displacement from a state of free equilibrium. This follows from an observation of CAPRIOLI[2], which may be extended to the case of large displacements and from which follows the existence of the elastic potential.

2. From HELMHOLTZ's postulate regarding the specific heat it follows that one may calculate T from (2.24, 2). Let U' be the function of ε_{rs}, E and $P*$ obtained by putting $T(\varepsilon_{pq}, E; P*)$ in the expression for the internal energy U. According to (2.21) there results

$$Y_{rs} = - \frac{\partial U'}{\partial \varepsilon_{rs}}. \tag{2.26}$$

From (2.23) follows

$$\delta * l^{(i)} = - \delta J_T = - \delta U'_E , \tag{2.27}$$

where δJ_T, $\delta U'_E$ are the variations of J and U' corresponding to variations of ε_{rs} only, T or E remaining constant. (2.24, 1) express Y_{rs} in terms of the ε_{pq} alone, for any isothermal transformation and in the case when

---

[1] An interesting definition of thermal equilibrium has been given recently by COLEMAN and NOLL. That definition is based on a *local inequality* regarding a kind of free energy per unit mass of the local state. The stress-strain relations are a corollary of the condition of equilibrium.

[2] See also UDESCHINI.

J is the sum of a function of the ε_{pq} only and of another function of T alone. Also for an adiabatic transformation (2.26) expresses Y_{rs} in terms of the ε_{pq} alone. In each of these cases equations (2.12), (2.13) [or (2.16)] depend upon the three unknown u_r only.

3. Let C' be a configuration characterized by some values of ε_{rs}, T', E', and let C'' be a configuration which is characterized by values ε_{rs}'' of the strain and which may be reached from C' by an isothermal or adiabatic displacement.

Let $dC*l_i^{(T)}$ and $dC*l_i^{(E)}$ be the work of the internal forces for the element $dC*$ corresponding to the two different types of displacement. Then $l_i^{(E)} < l_i^{(T)}$ [SIGNORINI, 3, p. 125].

4. Let φ_i be three parameters by which it is possible to represent the directions of the principal axes of strain, T_B at $P*$, and let i_i be the unit vectors of T_B; e. g., φ_i may be the three Euler angles. Let E_i be the principal components of strain. Put

$$\omega_r = \frac{1}{2} i_l \times \frac{\partial i^l}{\partial \varphi_r}, \qquad (2.28)$$

let ω_{rs} be the coordinates of ω_r and $y_{rs}^{(B)}$ the components of stress with reference to T_B. Then [SIGNORINI, 3, p. 128]

$$Y_{rs}^{(B)} = -\frac{\partial J^{(B)}}{\partial E_r}, \quad \sum_l Y_{l+1 l+2}^{(B)} \omega_{rl} (E_{l+2} - E_{l+1}) = -\frac{1}{2} \frac{\partial J^{(B)}}{\partial \varphi_r}, \quad (2.29)$$

where $J^{(B)} (E_i, \varphi_p, T; P*)$ is the function of E_i and φ_p obtained from $J (\varepsilon_{rs}, T; P*)$ by substituting for the ε_{rs} their expressions in terms of E_i, φ_p.

Chapter III

Isotropic bodies — Thermodynamic potential

§ 1. Isotropic bodies

1. A very difficult problem of the theory of thermoelastic transformations is the one of determining the nature of the thermodynamic potential, J. This problem is both theoretical and experimental: first of all theoretical, since it is necessary to set up by analytical, geometrical, kinematical, physical, etc., considerations some kind of dependence of J upon fundamental parameters; then experimental, since only experience may establish or overturn a proposed scheme. Also, experience may suggest modifications. Hooke's Law, which implies identification of J with a quadratic form in the linearized ε_{rs} and in T, represents a solution

of the problem in first approximation, but it fails if the deformations are not sufficiently small, even though they remain in the elastic range. Nor may it be said, for certainly it is not true, that the same form of thermodynamic potential is valid for every elastic body in the case of large strain.

The problem of finding the form of the function J, while difficult, certainly becomes simpler if one considers bodies whose nature is such as to impose upon J special properties, symmetry, etc. I shall consider only the case of isotropic bodies, where greater advances have been made.

Let us suppose, then, that the body is *isotropic in C**. This means that at least with respect to one reference configuration the form of the thermodynamic potential does not depend upon the frame of reference. It is well known that a necessary and sufficient condition for this isotropy is that J depends upon ε_{rs} only through the principal invariants of deformation. These are

$$I_1 = \varepsilon_r^{\cdot r}, \quad I_2 = \sum_r (\varepsilon_{rr} \varepsilon_{r+1 r+1} - \varepsilon_{rr+1}^2), \quad I_3 = ||\varepsilon_{rs}||. \qquad (3.1)$$

When referred to the principal axes of strain at P^*, (3.1) become

$$I_1 = E_r^{\cdot r}, \quad I_2 = \sum_r E_r E_{r+1}, \quad I_3 = E_1 E_2 E_3. \qquad (3.2)$$

where $E_r > -\dfrac{1}{2}$ are the principal components of strain at P^*. Another invariant which is often used is

$$D = \sqrt{1 + 2 I_1 + 4 I_2 + 8 I_3} = \sqrt{(1 + 2E_1)(1 + 2E_2)(1 + 2E_3)}. \qquad (3.3)$$

If the body is isotropic in C^*, the only case that will be considered in this chapter, J must be regarded as a function only of temperature T in the typical state, of temperature τ in C^* and of I_1, I_2, I_3 or, equivalently, of I_1, I_2, D, with coefficients which may depend upon the coordinates if the body is not homogeneous. I shall write $J(I_1, I_2, I_3; T, \tau)$ and $J^{(D)}(I_1, I_2, D; T, \tau)$ or more simply J and $J^{(D)}$, accordingly as J is taken as a function of I_1, I_2, I_3 or of I_1, I_2, D.

Except in the case of a perfect fluid, the property of isotropy depends upon the configuration, and the following theorem holds [SIGNORINI, 3]: *If the body is isotropic in C* and $J(I_1, I_2, D)$ does not reduce to a function of D only, a deformation from C* maintains the isotropy if and only if it preserves angles [conformal displacement].*

The body being supposed isotropic in C^*, let it be referred to the principal axes of strain at P^*. In this case J depends upon E_1, E_2, E_3 [by means of I_1, I_2, D] but not upon the parameters $\varphi_1, \varphi_2, \varphi_3$ (see Chap. II, § 3], and the $Y_{rs}^{(B)}$ are all equal to zero for $r \neq s$. This means that the principal axes of strain coincide with the principal axes of the Lagrangean tensor of stress, Y_{rs}. This property is characteristic for isotropy in C^* [GASPARINI].

2. From (3.1) follows

$$\frac{\partial I_1}{\partial \varepsilon_{rs}} = \delta_{rs}, \qquad \frac{\partial I_2}{\partial \varepsilon_{rs}} = (I_1 + \varepsilon_{rs}) \delta_{rs} - 2\varepsilon_{rs},$$

$$\frac{\partial I_3}{\partial \varepsilon_{rs}} = (2 - \delta_{rs}) (\varepsilon_{r+1\,s+1}\, \varepsilon_{r+2\,s+2} - \varepsilon_{r+1\,s+2}\, \varepsilon_{r+2\,s+1}). \qquad (3.4)$$

Then, since from (2.24, 1) it follows that

$$Y_{rs} = - \sum_i \frac{\partial J}{\partial I_i} \frac{\partial I_i}{\partial \varepsilon_{rs}}, \qquad (3.5)$$

one has

$$Y_{rs} = - \delta_{rs} \left(\frac{\partial J}{\partial I_1} + I_1 \frac{\partial J}{\partial I_2} \right) + (2 - \delta_{rs}) \frac{\partial J}{\partial I_2} \varepsilon_{rs}$$

$$+ (\delta_{rs} - 2) \frac{\partial J}{\partial I_3} (\varepsilon_{r+1s+1}\, \varepsilon_{r+2s+2} - \varepsilon_{r+1s+2}\, \varepsilon_{r+2s+1}). \qquad (3.6)$$

The Y_{rs} are the Lagrangean components of stress. Their principal interest consists in the fact that problems of integration and proof of several theorems may become easier if the Lagrangean point of view is used.

3. Doubtless the Eulerian components of stress X_{rs} are of greater interest for the knowledge of the actual stress in the body. But, while the Y_{rs}, according to (3.6), are polynomials of second degree in ε_{pq} with coefficients depending only on the principal invariants and the temperature, no analogous property holds for X_{rs}; in view of (3.6), (2.10) imply a real dependence of X_{rs} upon the local rotation as well, that is, upon ϱ_{rs} [see (1.27)].
    Specifically, putting

$$v_{rs} = \frac{1}{2} (u_{r,s} + u_{s,r} + u_{r,l}\, u_s^{;l}), \qquad (3.7)$$

one finds

$$X_{rs} = - \frac{1}{D} [l\delta_{rs} + 2mv_{rs} + nv_{rt}\, v_{.s}^t], \qquad (3.8)$$

where[1]

$$l = D \frac{\partial J^{(D)}}{\partial D} + \frac{\partial J^{(D)}}{\partial I_1} + I_1 \frac{\partial J^{(D)}}{\partial I_2},$$

$$m = \frac{\partial J^{(D)}}{\partial I_1} + I_1 \frac{\partial J^{(D)}}{\partial I_2} - \frac{1}{2} \frac{\partial J^{(D)}}{\partial I_2}, \qquad\qquad (3.9)$$

$$n = -2 \frac{\partial J^{(D)}}{\partial I_2}.$$

[1] The relations (3.8), (3.9) are the scalar expressions for the homographic ones given by SIGNORINI [3].

(3.8), (3.9) are equivalent to FINGER's relations, which may be put in the following form: let

$$g_{ih} = \delta_{ih} + 2v_{ih}, \quad \bar{I}_1 = g_s^{\cdot s}, \quad \bar{I}_2 = G_s^{\cdot s}, \quad \bar{I}_3 = \|g_{ih}\|, \quad (3.10)$$

where G_{ih} is the cofactor of g_{ih}, $\bar{I}_1$, $\bar{I}_2$, $\bar{I}_3$ are the first three invariants of g_{ih}; it is easy to recognize that $\bar{I}_i(g) = I_i(\varepsilon)$. Considering J as a function of $\bar{I}_1$, $\bar{I}_2$, $\bar{I}_3$, one has[1]

$$X_{rs} = -\frac{2}{\sqrt{\bar{I}_3}} \left[\delta_{rs} \frac{\partial J}{\partial \bar{I}_1} - G_{rs} \frac{\partial J}{\partial \bar{I}_2} + \delta_{rs} \left(\frac{\partial J}{\partial \bar{I}_2} \bar{I}_2 + \frac{\partial J}{\partial \bar{I}_3} \bar{I}_3 \right) \right]. \quad (3.11)$$

Several other forms of the relations by which X_{rs} are derived from a potential function were obtained long ago by several authors, as KIRCHHOFF, THOMSON, BOUSSINESQ [1], COSSERAT, etc., but it is of no present importance to recall them.

Relations analogous to (3.8) hold for isotropic bodies in more general cases than the elastic one, if the hypothesis is made that the planes of the circular sections of the indicator quadrics of the matrix $|\delta_{rs} + 2\varepsilon_{rs}|$ and of the Lagrangean stress deviator are the same [and besides, the axes of symmetry are the same] [MANACORDA, 1].

4. Since the expressions (3.8) for X_{rs} depend upon the local rotation, it may be convenient to express X_{rs} by means of the *inverse displacement*, that is, the displacement from C to C^*. Plainly, the components of that displacement are $-u_r$, and the corresponding components of strain ε'_{rr}, $2\varepsilon'_{r+1r+2}$ are

$$\varepsilon'_{rs} = -\frac{1}{2} \left[\frac{\partial u_r}{\partial x_s} + \frac{\partial u_s}{\partial x_r} - \sum_i \frac{\partial u_i}{\partial x_r} \frac{\partial u_i}{\partial x_s} \right]. \quad (3.12)$$

One may demonstrate that the X_{rs} as functions of the inverse displacement do not depend upon its local rotation. Specifically, if I'_i, D' denote the invariants analogous to those expressed by (3.1), (3.3), it is possible to show that a one-to-one correspondence exists between I_1, I_2, I_3 and I'_1, I'_2, I'_3, and that

$$D' = \frac{1}{D}. \quad (3.13)$$

Further, if $J'^{(D)}$ denotes the thermodynamic potential expressed as a function of I'_1, I'_2, D', from (2.10), (3.5) it follows[2] that

$$X_{rs} = l' \delta_{rs} + 2m' \varepsilon'_{rs} + n' \varepsilon'^{\cdot t}_r \varepsilon'_{ts}, \quad (3.14)$$

---

[1] Due to a different definition of $X_{r,s}$, the sign of the right-hand side of (3.11) is the opposite of that in FINGER's formula.

[2] Equalities (3.14), (3.15) are scalar expressions of the homographic ones given by SIGNORINI [3].

where

$$l' = k^* D' \left[D' \frac{\partial J'^{(D')}}{\partial D'} + \frac{\partial J'^{(D')}}{\partial I_1'} + I_1' \frac{\partial J'^{(D')}}{\partial I_2'} \right],$$

$$m' = k^* D' \left[\frac{\partial J'^{(D')}}{\partial I_1'} + I_1' \frac{\partial J'^{(D')}}{\partial I_2'} - \frac{1}{2} \frac{\partial J'^{(D')}}{\partial I_2'} \right], \qquad (3.15)$$

$$n' = - 2 k^* D' \frac{\partial J'^{(D')}}{\partial I_2'} .$$

§ 2. Principal properties of the thermodynamic potential

Doubtless the most interesting problem of thermoelasticity is that of determining the thermodynamic potential. Experience may give a quantitative check, but the mathematical theory must first suggest a hypothesis on its analytic form; A. SIGNORINI during a recent lecture at the Milan Institute of Technology has rightly said that the problem should be given to experimentalists as late as possible.

That is, it is best to determine first as many properties as possible that the thermodynamic potential must necessarily satisfy, so as to simplify the extremely difficult problem of finding its form, at least for an assigned body.

Above all, the *basic property* of elastic bodies must be kept in mind: *configurations of free equilibrium exist such that the work done by the internal contact forces is negative for any isothermal non-rigid displacement starting from them, when the temperature τ lies between τ_1 and $\tau_2 > \tau_1$.*

The displacement from a configuration of free equilibrium to another one is certainly a rigid displacement.

A. SIGNORINI [4] clearly has rendered explicit some properties which must be attributed to the thermodynamic potential $J^{(D)} (I_1, I_2, D; T, \tau)$ by virtue of the fact that it refers to a homogeneous, elastic body, isotropic in the reference state C^* of free equilibrium. By using those asserted properties it is possible to infer that $J^{(D)} (I_1, I_2, D; T, \tau)$ must satisfy the following conditions:

a) *Denote by k_τ^* the density in C^* at temperature τ, and by $W_\tau(I_1, I_2, D)$ the isothermal elastic potential at temperature τ, defined by*

$$W_\tau (I_1, I_2, D) = k_\tau [J^{(D)} (I_1, I_2, D; \tau, \tau) - J^{(D)} (0, 0, 1; \tau, \tau)] \quad (3.16)$$

and equal to zero in C^*; then

$$V_\tau = \int_{C^*} W_\tau (I_1, I_2, D) \, dC^* \geqslant 0, \qquad (3.17)$$

where the sign of equality is valid only for a rigid displacement. Property a) is an immediate consequence of the *basic property* of elastic bodies.

b) *Any state of free equilibrium is a natural state* [unstressed state]. Property b) is a consequence of the hypothesis of homogeneity. It follows also from the hypothesis of isotropy independently of the former. Therefore, in C^* we have

$$X_j^{(\tau)} = Y_j^{(0)} = -\left(\frac{\partial J^{(D)}}{\partial \varepsilon_j}\right)_{T=\tau}^{(0)} = -\left(\frac{\partial W_\tau}{\partial \varepsilon_j}\right)^{(0)} = 0. \qquad (3.18)$$

c) *For any non-rigid displacement starting from C^*,*

$$W_\tau > 0. \qquad (3.19)$$

Property c) is an immediate consequence of the hypothesis of homogeneity and of (3.17).

d) *The thermodynamic potential $J^{(D)}$ and the elastic potential W_τ may not depend upon the components of strain through the invariant D alone.* Property d) is a consequence of (3.16), (3.17).

e) *A transformation from C^* in which the temperature varies from τ to $\tau' \neq \tau$ preserves the homogeneity and isotropy if and only if it is a similitude.*

Property e) is a consequence of d) [SIGNORINI, 4, p. 9].

f) *The analytical form of $J^{(D)}$ must not vary in passing from one natural state to another one with different temperature.*

In other words, when τ belongs to a certain interval $\tau_1 \leftrightarrow \tau_2$, the analytical structure of $J^{(D)}$ must be invariant with respect to the natural reference state, as physical intuition suggests, so that it is not possible to give privileged position to some special configuration of natural equilibrium.

Therefore, it is necessary that

$$J^{(D)}(I_1^{(\tau)}, I_2^{(\tau)}, D^{(\tau)}; T, \tau) = J^{(D)}(I_1^{(\tau')}, I_2^{(\tau')}, D^{(\tau')}; T, \tau') \qquad (3.20)$$

for τ, τ' belonging to the interval $\tau_1 \leftrightarrow \tau_2$.

g) *The isothermal elastic potential W_τ tends to infinity if and only if at least one of the principal components of strain tends to* $-\frac{1}{2}$ *or to infinity.*

h) *In uniform traction or pressure the linear coefficient of dilatation is positive or negative respectively.*

i) *In simple extension of a cylindrical body the stress resultant on a cross section is an increasing function of the elongation per unit length.*

l) *$J^{(D)}$ satisfies the postulate of Helmholtz.*

§ 3. A property of the thermodynamic potential

1. Property f) of the thermodynamic potential $J^{(D)}$, remarked some time ago by SIGNORINI [4], restricts the functional dependence of $J^{(D)}$ upon its arguments. In fact, if, on the one hand, by property e) a deformation caused by variation of the temperature of C^* is a similitude,

on the other hand the form of the functional dependence of $J^{(D)}$ upon I_1, I_2, D changes.

SIGNORINI studied and solved a problem of this kind in regard to an interesting type of thermodynamic potential proposed by him, which I shall consider later [§ 4]; TOLOTTI [4] has taken up the general question and resolved it.

Specifically, TOLOTTI has determined the most general function $J^{(D)}(I_1, I_2, D; T, \tau)$ which satisfies property f), demonstrating that that property does not restrict the form of the elastic potential W_τ but implies that $J^{(D)}$ depends upon W_τ, the density k_τ in C^* and the specific heat at constant pressure $c_p^{(\tau)}$.

Let us consider two possible natural reference states $C_\tau^*, C_{\tau'}^*$, respectively, at temperatures $\tau, \tau' \neq \tau$ belonging to a certain interval $\tau_1 \leftrightarrow \tau_2$. By property e) the displacement $C_\tau^* \to C_{\tau'}^*$ is a similitude; let its characteristic quotient be $\dfrac{l_{\tau'}}{l_\tau}$.

It is easy to see that between the invariants $I_i^{(\tau)}, I_i^{(\tau')}$ corresponding to the deformations $C_\tau^* \to C$ and $C_{\tau'}^* \to C$ the following relations hold:

$$l_\tau^2 (3 + 2 I_1^{(\tau)}) = l_{\tau'}^2 (3 + 2 I_1^{(\tau')}),$$
$$l_\tau^4 (3 + 4 I_1^{(\tau)} + 4 I_2^{(\tau)}) = l_{\tau'}^4 (3 + 4 I_1^{(\tau')} + 4 I_2^{(\tau')}), \qquad (3.21)$$
$$l_\tau^3 D^{(\tau)} = l_{\tau'}^3 D^{(\tau')}.$$

The solutions of (3.20) are those functions $J^{(D)}$ which satisfy the equation

$$\frac{dJ^{(D)}}{d\tau} = \frac{\partial J^{(D)}}{\partial \tau} + \frac{\partial J^{(D)}}{\partial I_1^{(\tau)}} \frac{dI_1^{(\tau)}}{d\tau} + \frac{\partial J^{(D)}}{\partial I_2^{(\tau)}} \frac{dI_2^{(\tau)}}{d\tau} + \frac{\partial J^{(D)}}{\partial D^{(\tau)}} \frac{dD^{(\tau)}}{\partial \tau} = 0, \qquad (3.22)$$

the general integral of which is an arbitrary function of

$$\left. \begin{aligned}
\bar{I}_1^{(\tau)} &= l_\tau^2 (3 + 2 I_1^{(\tau)}), \\
\bar{I}_2^{(\tau)} &= l_\tau^4 (3 + 4 I_1^{(\tau)} + 4 I_2^{(\tau)}), \\
\bar{D}^{(\tau)} &= l_\tau^3 D^{(\tau)}.
\end{aligned} \right\} \qquad (3.23)$$

One concludes that the most general thermodynamic potential which satisfies property f) is an arbitrary function of T and of $\bar{I}_1^{(\tau)}, \bar{I}_2^{(\tau)}, \bar{D}^{(\tau)}$, depending upon τ only through l_τ.

From (3.21), by identifying τ' with the temperature T in C, it follows that,

$$\left. \begin{aligned}
I_1^{(T)} &= \frac{l_\tau^2}{l_T^2} \left(\frac{3}{2} + I_1^{(\tau)} \right) - \frac{3}{2}, \\
I_2^{(T)} &= \frac{l_\tau^4}{l_T^4} \left(\frac{3}{4} + I_1^{(\tau)} + I_2^{(\tau)} \right) - \frac{l_\tau^2}{l_T^2} \left(\frac{3}{2} + I_1^{(\tau)} \right) + \frac{3}{4}, \\
D^{(T)} &= \frac{l_\tau^3}{l_T^3} D^{(\tau)},
\end{aligned} \right\} \qquad (3.24)$$

which, by (3.23), may be written

$$
\left.\begin{aligned}
I_1^{(T)} &= \frac{I_1^{(\tau)}}{2\,l_T^2} - \frac{3}{2}\,, \\[2mm]
I_2^{(T)} &= \frac{I_2^{(\tau)}}{4\,l_T^4} - \frac{I_1^{(\tau)}}{2\,l_T^2} + \frac{3}{4}\,, \\[2mm]
D^{(T)} &= \frac{\bar{D}^{(\tau)}}{l_T^3}\,.
\end{aligned}\right\}
\tag{3.25}
$$

Since $J^{(D)}$ must be equal to a function of $\bar{I}_1^{(\tau)}$, $\bar{I}_2^{(\tau)}$, $\bar{D}^{(\tau)}$, according to (3.25) one may put

$$
J^{(D)}(I_1^{(\tau)}, I_2^{(\tau)}, D^{(\tau)}; T, \tau) = \Phi(I_1^{(T)}, I_2^{(T)}, D^{(T)}; T).
\tag{3.26}
$$

From (3.16), (3.26) follows

$$
\frac{1}{k_\tau}\, W_\tau(I_1^{(\tau)}, I_2^{(\tau)}, D^{(\tau)}) = \Phi(I_1^{(\tau)}, I_2^{(\tau)}, D^{(\tau)}, \tau) - \Phi(0, 0, 1, \tau).
\tag{3.27}
$$

Then (3.26) becomes

$$
J^{(D)}(I_1^{(\tau)}, I_2^{(\tau)}, D^{(\tau)}; T, \tau) = \frac{1}{k_T}\, W_T(I_1^{(T)}, I_2^{(T)}, D^{(T)}) + \varphi(T),
\tag{3.28}
$$

where

$$
\varphi(T) = J^{(D)}(0, 0, 1; T, T).
\tag{3.29}
$$

To determine $\varphi(T)$ it must be observed first of all that from (3.28) follows

$$
\frac{d\varphi}{dT} = \frac{\partial J^{(D)}}{\partial T} + \frac{1}{k_T^2}\frac{dk_T}{dT}\, W_T(I_1^{(T)}, I_2^{(T)}, D^{(T)}) - \frac{1}{k_T}\left[\frac{\partial W_T}{\partial I_1^{(T)}}\frac{\partial I_1^{(T)}}{\partial T} + \right.
$$
$$
\left. + \frac{\partial W_T}{\partial I_2^{(T)}}\frac{\partial I_2^{(T)}}{\partial T} + \frac{\partial W_T}{\partial D^{(T)}}\frac{\partial D^{(T)}}{\partial T}\right].
\tag{3.30}
$$

On the other hand, denoting by δ_τ the coefficient of thermal dilatation in C_τ^* at temperature τ, from (3.24) follows

$$
\left(\frac{\partial I_1^{(T)}}{\partial T}\right)_{\varepsilon^{(\tau)}=0,\,T=\tau} = \left(\frac{\partial D^{(T)}}{\partial T}\right)_{\varepsilon^{(\tau)}=0,\,T=\tau}
$$
$$
= -3\delta_\tau, \quad \left(\frac{\partial I_2^{(T)}}{\partial T}\right)_{\varepsilon^{(\tau)}=0,\,T=\tau} = 0,
\tag{3.31}
$$

while from (3.16) one deduces

$$
[W_\tau]_{\varepsilon^{(\tau)}=0} = \left[\frac{\partial W_\tau}{\partial \tau}\right]_{\varepsilon^{(\tau)}=0} = 0.
\tag{3.32}
$$

According to (3.31), (3.32), from (3.30) it follows that

$$
\left[\frac{d\varphi}{dT}\right]_{T=\tau} = \left[\frac{\partial J}{\partial T}\right]_{\varepsilon^{(\tau)}=0,\,T=\tau} + 3\frac{\delta_\tau}{k_\tau}\left[\frac{\partial W_\tau}{\partial I_1^{(\tau)}} + \frac{\partial W_\tau}{\partial D^{(\tau)}}\right]_{\varepsilon^{(\tau)}=0}.
\tag{3.33}
$$

Then, it is sufficient to bear in mind (3.8), (3.9) in order to recognize that

$$\left[\frac{\partial W_\tau}{\partial I_j^{(\tau)}} + \frac{\partial W_\tau}{\partial D^{(\tau)}}\right]_{\varepsilon^{(\tau)}=0} = [l]_{\varepsilon^{(\tau)}=0} = [X_{rr}]_{\varepsilon^{(\tau)}=0} = 0, \qquad (3.34)$$

since C_τ^* is a state of natural equilibrium. Therefore, from (2.24, 2), (3.33) it follows that

$$\left[\frac{d\varphi}{dT}\right]_{T=\tau} = \left[\frac{\partial J^{(D)}}{\partial T}\right]_{\varepsilon^{(\tau)}=0,\, T=\tau} = -eE_\tau, \qquad (3.35)$$

where E_τ denotes the entropy in C_τ^* at temperature τ. If $c_p^{(\tau)}$ is the specific heat at constant pressure,

$$E_\tau = \int_{\tau_0}^{\tau} c_p^{(\tau)} \frac{d\tau}{\tau}, \qquad (3.36)$$

and, finally, from (3.35), (3.36) it follows that

$$\varphi(T) = -e\int_{T_0}^{T} d\tau' \int_{\tau_0}^{\tau'} c_p^{(\tau'')} \frac{d\tau''}{\tau''} \qquad (3.37)$$

which, according to (3.28), determines $J^{(D)}$ in terms of the isothermal potential W_τ to within an inessential linear function of T.

2. One finds an analogous result for the inverse displacement $C \to C_\tau^*$ if one requires that $J^{(D')}(I_1', I_2', D'; T, \tau)$ exhibit property f). Specifically, for given isothermal potential $W_T(I_1', I_2', D')$ the expression for $J^{(D')}$ is

$$J^{(D')}(I_1'^{(\tau)}, I_2'^{(\tau)}, D'^{(\tau)}; T, \tau) = \frac{1}{kT} W_T(I_1'^{(T)}, I_2'^{(T)}, D'^{(T)}) + \varphi(T), \quad (3.38)$$

where $\varphi(T)$ is still given by (3.37), and where

$$
\left.
\begin{aligned}
I_1'^{(T)} &= \frac{l_T^2}{l_\tau^2}\left(\frac{3}{2} + I_1'^{(\tau)}\right) - \frac{3}{2}, \\[4pt]
I_2'^{(T)} &= \frac{l_T^4}{l_\tau^4}\left(\frac{3}{4} + I_1'^{(\tau)} + I_2'^{(\tau)}\right) - \frac{l_T^2}{l_\tau^2}\left(\frac{3}{2} + I_1'^{(\tau)}\right) + \frac{3}{4}, \\[4pt]
D'^{(T)} &= \frac{l_T^3}{l_\tau^3} D'^{(\tau)}.
\end{aligned}
\right\}
\qquad (3.39)
$$

§ 4. Elasticity of second grade. Signorini's thermodynamic potential

1. It has been seen that determination of the thermodynamic potential depends upon that of the isothermal potential $W_\tau(I_1^{(\tau)}, I_2^{(\tau)}, D^{(\tau)})$. The experimental determination of W_τ is practically impossible without some preliminary indication of its analytical form, the more so in that experiments on traction and pressure, which are the ones more easily

made, determine W_τ only in a neighborhood of a line in the threedimensional region $(I_1^{(\tau)}, I_2^{(\tau)}, D^{(\tau)})$ where W_τ is defined [SIGNORINI, 4, p. 27].

Several authors have proposed different forms for the dependence of stress upon strain, but often the results are not free from objection due to their approximate character or other reasons. It is not within my scope to recall here the theories of this kind due to DE SAINT-VENANT [1, 2], VOIGT [2, 3], KÖTTER, SOUTHWELL, HENCKY [1], SETH, MURNAGHAN [1, 2], ZVOLINSKI and RIZ.

Those theories have been summarized by TRUESDELL [2, 3], with some critical observations. More recent results in this field have been obtained by NOVOZILOV [2], whose stress-strain relations involve three moduli: a generalized bulk modulus, a generalized shear modulus and a phase angle between the deviators of stress and strain; by BAKER and ERICKSEN, who suppose that the tension is largest (smallest) in the direction of the maximum (minimum) elongation; by ERICKSEN [1] who supposes that X_{rs} depends only upon $x^{i,h}$ without there necessarily existing a strain-energy; and by other authors, such as HANIN and REINER, CSONKA, CHU BOA-TEH, SMITH and RIVLIN, etc.

Granted the convenience of referring to the inverse displacement $C \to C^*$, the simplest possible hypothesis consists in assuming that the principal tensions B_1, B_2, B_3 are linear functions of the principal components of strain of the inverse displacement, E_s'. However, it will be shown soon that this hypothesis may not be accepted, since it is in contradiction with property c), expressed by (3.19). Therefore, the most natural hypothesis, proposed and studied by SIGNORINI [4, Chap. II] is that *each principal tension be a quadratic function of the principal components of strain of the inverse displacement*.

Thus emerges an exact quadratic theory which represents an important advance in the field. The corresponding expression of the thermodynamic potential satisfies all principal properties of § 2, even though it involves only a small number of coefficients.

The expression of $J'^{(D')}$, established by SIGNORINI in the most general form, may be obtained readily by the following considerations. Plainly the condition that the B_r be quadratic functions of E_s' is equivalent to requiring that $J'^{(D')}$ makes l' in (3.15) a quadratic function, m' a linear function of I_1' and n' independent of I_1', I_2', D'. The most general functions of this kind such that (3.15) are integrable are

$$\left.\begin{array}{l} l' = p + (2\beta - \alpha + c)\, I_1' - \beta\, I_1'^{\,2} - c I_2', \\[2mm] m' = \alpha - \dfrac{c}{2} + (2\beta + c)\, I_1', \\[2mm] n' = -\, 2c, \end{array}\right\} \qquad (3.40)$$

where p, α, β, c depend upon τ and T only.

By (3.40) and integrating (3.15), one deduces

$$J'^{(D')} = \frac{1}{k_\tau D'}[-p + \alpha(I_1' + 1) + \beta I_1'^2 + cI_2'] - q \qquad (3.41)$$

where q depends upon τ and T only.

If the isothermal elastic potential of the inverse transformation is defined analogously to (3.16), from (3.41) follows

$$W_\tau'^{(D')}(I_1'^{(\tau)}, I_2'^{(\tau)}, D'^{(\tau)}) = \frac{1}{D'^{(\tau)}}[\alpha(\tau, \tau)(I_1'^{(\tau)} + 1) + \beta(\tau, \tau) I_1'^{(\tau)^2} +$$

$$+ c(\tau, \tau) I_2'^{(\tau)}] - \alpha(\tau, \tau) - p(\tau, \tau)\left[\frac{1}{D'^{(\tau)}} - 1\right]. \qquad (3.42)$$

2. From (3.34), valid also with reference to the inverse displacement, and from (3.42), one may show that

$$p(\tau, \tau) = 0 ; \qquad (3.43)$$

from (3.38), (3.42) it follows that

$$J'^{(D')} = \frac{1}{k_T D'^{(T)}}[\alpha(T, T)(I_1'^{(T)} + 1) + \beta(T, T) I_1'^{(T)^2} +$$

$$+ c(T, T) I_2'^{(T)}] + \varphi(T) - \frac{\alpha(T, T)}{k_T}. \qquad (3.44)$$

In virtue of (3.39), (3.44) becomes

$$J'^{(D')} = \frac{1}{k_T D'^{(T)}}\left\{\frac{l_T^2 - l_\tau^2}{2 l_\tau^4}\left[\left(\alpha(T, T) - \frac{9}{2}\beta(T, T) - \frac{3}{2}c(T, T)\right)l_\tau^2 - \right.\right.$$

$$- (3\beta(T, T) + c(T, T))\frac{l_T^2}{2}\right] + \frac{l_T^2}{l_\tau^4}[(\alpha(T, T) - 3\beta(T, T) -$$

$$- c(T, T))l_\tau^2 + (3\beta(T, T) + c(T, T))l_T^2](I_1'^{(\tau)} + 1) +$$

$$+ \frac{l_T^4}{l_\tau^2}\beta(T, T)I_1'^{(\tau)^2} + \frac{l_T^4}{l_\tau^4}c(T, T)I_2'^{(\tau)}\right\} + \varphi(T) - \frac{\alpha(T, T)}{k_T}. \qquad (3.45)$$

Let us put

$$\alpha(\tau, \tau) = \mu_\tau + \frac{c_\tau}{2}, \qquad \beta(\tau, \tau) = \frac{1}{2}\left(\lambda_\tau + \mu_\tau - \frac{c_\tau}{2}\right). \qquad (3.46)$$

Recalling (3.37) and the invariance of $k_\tau D'^{(\tau)}$, finally one finds that the thermodynamic potential is expressed by (3.41), where[1]

$$\left.\begin{array}{l}
p = \dfrac{l_T^2 - l_\tau^2}{8\, l_\tau^4}\left[3\left(\lambda_T + \mu_T + \dfrac{c_T}{6}\right)l_T^2 + \left(9\lambda_T + 5\mu_T - \dfrac{c_T}{2}\right)l_\tau^2\right], \\[2.5ex]
\alpha = \dfrac{l_T^2}{2\, l_\tau^4}\left[3\left(\lambda_T + \mu_T + \dfrac{c_T}{6}\right)l_T^2 - \left(3\lambda_T + \mu_T - \dfrac{c_T}{2}\right)l_\tau^2\right], \\[2.5ex]
\beta = \dfrac{l_T^4}{2\, l_\tau^4}\left(\lambda_T + \mu_T - \dfrac{c_T}{2}\right), \\[2.5ex]
c = \dfrac{l_T^4}{l_\tau^2}\, c_T,
\end{array}\right\} \qquad (3.47)$$

---

[1] For a relation among isothermal or adiabatic elastic moduli at various temperatures see BRILLOUIN.

$$q = \frac{1}{2 k_T} (2\mu_T + c_T) + e \int_{T_0}^{T} d\tau' \int_{\tau_0}^{\tau'} c_p^{(\tau'')} \frac{d\tau''}{\tau''}. \tag{3.48}$$

The corresponding isothermal potential is

$$W_\tau'^{(D')} = \frac{1}{D'^{(\tau)}} \left[\left(\mu_\tau + \frac{c_\tau}{2} \right) (I_1'^{(\tau)} + 1) + \frac{1}{2} \left(\lambda_\tau + \mu_\tau - \frac{c_\tau}{2} \right) I_1'^{(\tau)^2} + \right.$$

$$\left. + c_\tau I_2'^{(\tau)} \right] - \left(\mu_\tau + \frac{c_\tau}{2} \right). \tag{3.49}$$

3. Since $W_\tau = 0$ in C_τ^*, property c) implies that the quadratic form

$$W_\tau^{(2)} = \frac{1}{2} \sum_{j, l=1}^{6} \left(\frac{\partial^2 W_\tau}{\partial \varepsilon_j \partial \varepsilon_l} \right)_{\substack{(\tau) \\ \varepsilon_{l=0}}} \varepsilon_j \varepsilon_l \tag{3.50}$$

is positive-definite. It is easily shown that

$$W_\tau^{(2)} = \frac{1}{2} [(\lambda_\tau + 2\mu_\tau) I_1'^{(\tau)^2} - 4\mu_\tau I_2'^{(\tau)}]. \tag{3.51}$$

Then follows the necessity of the well known conditions for the Lamé coefficients

$$\mu_\tau > 0, \qquad 3\lambda_\tau + 2\mu_\tau > 0. \tag{3.52}$$

But (3.52) are only necessary conditions that W_τ be positive-definite.
Signorini [4] established (3.47), (3.48) directly supposing

$$c_\tau = c_T = c = 0. \tag{3.53}$$

In view of (3.14), (3.15), (3.40), (3.47), it is easy to see that hypothesis (3.53) is equivalent to requiring that X_{rr} shall not depend upon ε_{rs} $(r \neq s)$.

Later Tolotti [3] considered the general case of quadratic elasticity, supposing $c \neq 0$.

If (3.53) are valid, to verify property c) it is necessary and sufficient that

$$\mu_\tau > 0, \qquad 9\lambda_\tau + 5\mu_\tau \geqslant 0, \tag{3.54}$$

which are more restrictive than (3.52).

To justify (3.54) let us observe that

$$x_\tau' = 1 + 2 E_\tau' > 0, \tag{3.55}$$

while

$$18 D' \leqslant (3 + 2 I_1') (6 + 2 I_1'), \tag{3.56}$$

as may easily be shown. From (3.56) it follows that $D' \leqslant 1$ in any deformation which makes $I_1' = 0$. Then, on the supposition that $c_\tau = 0$, by (3.49) the necessity of condition (3.54, 1) is evident. Analogously for (3.54, 2): it is sufficient to observe that if E_r' approaches $-\frac{1}{2}$, D' and $3 + 2 I_1'$ approach 0. To show the sufficiency of conditions (3.54) let us write (3.56) in the form

$$(18)^2 x_1' x_2' x_3' < (\textstyle\sum_r x_r')^2 (3 + \textstyle\sum_r x_r')^2 \tag{3.57}$$

and observe that

$$x_1' x_2' x_3' < \frac{1}{27} (\textstyle\sum_r x_r')^3 , \qquad \textstyle\sum_r x_r' < \frac{1}{12} (3 + \textstyle\sum_r x_r')^2 . \tag{3.58}$$

Then the sufficiency of (3.54) becomes evident if one observes that when $c_\tau = 0$, (3.49) may be put into the form

$$W_\tau'^{(D')} = \mu_\tau \left[\frac{(3 + 2 I_1'^{(\tau)}) (6 + 2 I_1'^{(\tau)})}{18 D'^{(\tau)}} - 1 \right] + \frac{9\lambda_\tau + 5\mu_\tau}{18 D'^{(\tau)}} I_1'^{(\tau)^2} . \tag{3.59}$$

(3.54) are also necessary and sufficient conditions in order that $W_\tau'^{(D')}$ become infinite only when at least one of the E_r' tends towards -1 or ∞, a physically desirable property.

4. Since the thermodynamic potential $J'^{(D')}$ must satisfy the Helmholtz postulate [property l)],

$$c_v = - \frac{T}{e} \frac{\partial^2 J}{\partial T^2} > 0 \tag{3.60}$$

where c_v is the specific heat at constant configuration. From (3.41), (3.47), (3.48) for $c = c_T = 0$, $l_T = l_\tau = \text{const.}$, $k_T = \text{const.}$, one infers that

$$-\frac{1}{D'(T)} \left[\frac{d^2 \mu_T}{dT^2} (I_1'^{(T)} + 1) + \frac{I_1'^{(T)^2}}{2} \frac{d^2}{dT^2} (\mu_T + \lambda_T) \right] +$$

$$+ \frac{d^2 \mu_T}{dT^2} + e k_T \frac{c_p^{(T)}}{T} > 0 . \tag{3.61}$$

Noticing the analogy between the expression formed by all terms of the first member of (3.61) except the last one and the expression (3.49) of W_τ for $c_\tau = 0$, one finds that property l) is satisfied if and only if

$$\frac{d^2 \mu_T}{dT^2} \leqslant 0 , \qquad 9 \frac{d^2 \lambda_T}{dT^2} + 5 \frac{d^2 \mu_T}{dT^2} \leqslant 0 . \tag{3.62}$$

5. It is now seen at a glance that a theory in which the B_r are linear functions of E_s' cannot be correct [SIGNORINI, 4]. In fact, under such a hypothesis, from (3.14), (3.15), (3.40) it follows that $n' = \beta = c = 0$ and that, consequently, (3.42) reduces to

$$W_\tau = \alpha \left[\frac{I_1' + 1}{D'} - 1 \right] , \tag{3.63}$$

which cannot satisfy property c). To see this, consider a deformation for which $I'_1 = 0$ and then $D' \leqslant 1$; then α must be positive; then for sufficiently large values of E'_r expression (3.63) yields a negative value for W_r.

6. SIGNORINI applies his quadratic theory to several isothermal problems: uniform traction or pressure, simple extension, cylindrical body subjected to an uniform pressure only on the lateral surface, bending and traction of a rectangular plate. The results are very interesting since it is possible to recognize that the theory exhibits also properties h), i), even though the expression of W_r involves only two coefficients. Without going into detail, I only observe here that in the problem of simple extension, according to SIGNORINI's theory, for the classical relation by which the coefficient of lateral contraction [Poisson's ratio] is defined, the following one is to be substituted:

$$e'' = \frac{1 - \sqrt{1 + 4\chi_\tau e' + e'^4}}{2} = -\chi_\tau e' - \frac{1 - 4\chi_\tau^2}{4} e'^2 + \ldots - \qquad (3.64)$$

where e' and e'' are connected with the coefficients of linear dilatation δ', δ'' in the longitudinal and transversal directions by the formulae

$$e' = \frac{1}{2}\left[1 - \frac{1}{(1 + \delta')^2}\right], \qquad e'' = \frac{1}{2}\left[1 - \frac{1}{(1 + \delta'')^2}\right], \qquad (3.65)$$

and further

$$\chi_\tau = \frac{\lambda_\tau}{2(\lambda_\tau + \mu_\tau)}. \qquad (3.66)$$

TRUESDELL [2] derives a formula regarding the simple shear problem from SIGNORINI's theory[1].

§ 5. On the most general elasticity of second grade

1. The expression (3.41) for the thermodynamic potential is the most general one for $J'^{(D')}$ in the case when the Eulerian components of stress are supposed to be quadratic polynomials in ε'_{rs}. Even supposing $c = 0$, the corresponding theory gives interesting results, but, certainly, the possibility of using also the coefficient c broadens the theory, especially in regard to the study of problems of pure shear, torsion, etc. TOLOTTI [3] has demonstrated that c may be unequal to zero and has established the conditions which the coefficients in (3.41) must satisfy in order that the corresponding expression of $J'^{(D')}$ be acceptable as a thermodynamic potential.

---

[1] In [TRUESDELL, 2, 3] there is an accurate critical recapitulation of the works regarding the mechanics of continuous media up to 1953, provided with copious references. See also [TRUESDELL, 8].

The results of ToLOTTI are summarized in the following considerations. Putting

$$v_\tau = \frac{c_\tau}{2\,(\lambda_\tau + \mu_\tau)}\,,\qquad (3.67)$$

let us remember (3.66) and consider the plane of points $Q_\tau \equiv (v_\tau, \chi_\tau)$, referred to two perpendicular Cartesian axes.

In this plane let us consider the points

$$B \equiv (-3,\, -1)\,,\qquad C \equiv \left(1,\, -\frac{1}{2}\right),\qquad D \equiv (1,\, 0)\,,\qquad F \equiv \left(0,\, \frac{1}{2}\right),$$

$$G \equiv \left(-3,\, -\frac{1}{2}\right)\qquad\qquad (3.68)$$

and the parabolic arcs

$$(v_\tau - 4\,\chi_\tau)^2 - v_\tau - 4 = 0\,,\qquad -3 < v_\tau \leqslant 0\,,\qquad -\frac{1}{2} \leqslant \chi_\tau \leqslant \frac{1}{2}\,;\qquad (3.69)$$

$$4\,\chi_\tau^2 + v_\tau - 1 = 0\,,\qquad 0 \leqslant v_\tau \leqslant 1\,,\qquad \chi_\tau \geqslant 0\,,\qquad (3.70)$$

denoting by A the region included within these arcs and the segments GB, BC, CD and by f_a its boundary.

The following theorem holds: *a necessary and sufficient condition for property g) is that*

$$\lambda_\tau + \mu_\tau > 0\qquad\qquad (3.71)$$

and that the point $Q_\tau \equiv (v_\tau, \chi_\tau)$ belong to region A.

This condition is also sufficient that $W_\tau > 0$.

2. Let us consider the problem of uniform pressure. Denoting by δ the coefficient of linear dilatation which is constant, independent of the direction and connected to the common value E' of the E_τ by the relation

$$E'_\tau = \frac{1}{2}\left[\frac{1}{(1+\delta)^2} - 1\right] = E' > -\frac{1}{2}\,,\qquad (3.72)$$

from (3.14), (3.40), (3.47), (3.66), (3.67) one may show that

$$B_1 = B_2 = B_3 = (\lambda_\tau + \mu_\tau)\left[2\,(1 + \chi_\tau)\,E' + \frac{1}{2}\,(3 + v_\tau)\,E'^2\right].\qquad (3.73)$$

Recalling the above theorem and (3.73), one sees that B_i has the same sign as E', and, for $B_i > 0$ [the case of pressure], E' increases more slowly than B_i, in accordance with the experimental results [VOIGT, 2].

3. Now let us consider the problem of simple extension of a cylindrical body. If δ' and δ'' are the coefficients of dilatation along and normal to the axis, let e' and e'' still be defined by (3.65). Let T be the stress resultant on the cross-section, the area of which in the unstrained state is A^*.

From (3.14), (3.40), (3.47), (3.66), (3.67) one deduces that

$$\bar{T} = - \frac{A^* B_1}{1 - 2 e''} = (\lambda_\tau + \mu_\tau) \, A F (e') \, , \qquad B_2 = B_3 = 0 \, , \qquad (3.74)$$

where

$$F(e') = - (1 + 2 \chi_\tau) + (1 + \nu_\tau) \, e' + \frac{1 + 2 \chi_\tau + (1 - \nu_\tau) \, (e' - e'^2)}{\sqrt{1 + 4 \nu_\tau e' + (1 - \nu_\tau) \, e'^2}} . \quad (3.75)$$

It may be shown that $F(e')$ is a real and continuous function of e', if $-\infty < e' < \frac{1}{2}$.

Let us consider in the plane (ν_τ, χ_τ) the points

$$A \equiv (-3, -1) \, , \quad B \equiv \left(-1, -\frac{3}{4}\right), \quad C \equiv \left(1, -\frac{1}{2}\right), \qquad D \equiv (1, 0) \, ,$$

$$F \equiv \left(0, \frac{1}{2}\right) \qquad\qquad\qquad (3.76)$$

and the curvilinear arcs having the equations

$$(1 + \nu_\tau)^2 (5 - \nu_\tau + 8 \chi_\tau) = (1 - \nu_\tau + 4 \chi_\tau)^2, \quad -3 \leqslant \nu_\tau \leqslant 0, \quad -1 \leqslant \chi_\tau \leqslant \frac{1}{2},$$
$$(3.77)$$

$$(1 - \nu_\tau) \, (\nu_\tau^2 + 4 \nu_\tau - 1)^2 \, (5 - \nu_\tau + 8 \pi) = (1 + \nu_\tau)^4 \, (1 - \nu_\tau + 4 \chi_\tau)^2,$$

$$-3 \leqslant \nu_\tau \leqslant -1, \qquad -1 \leqslant \chi_\tau \leqslant -\frac{3}{4} . \qquad (3.78)$$

These arcs together with segments BC, CD and the parabolic arc (3.70) determine a curvilinear pentagon A'. The following theorem holds: *In order for $F(e')$ to be an increasing function of e' when e' varies from $-\infty$ to $\frac{1}{2}$ or—if one prefers—an increasing function of δ' when δ' varies from -1 to $+\infty$ [property i)], it is necessary and sufficient that (3.71) be satisfied and that the point $Q_\tau \equiv (\nu_\tau, \chi_\tau)$ belong to A'.*

The lateral contraction is now defined not by (3.64) but by

$$e'' = \frac{1 - \sqrt{1 + 4 \chi_\tau e' + (1 - \nu_\tau) \, e'^2}}{2} = - \chi_\tau e' - \frac{1 - 4 \chi_\tau^2 - \nu_\tau}{4} \, e'^2 + \dots.$$

$$(3.79)$$

§ 6. A new type of thermodynamic potential proposed by Tolotti

The expression of the thermodynamic potential proposed by SIGNO-RINI, like those of ALMANSI, HENCKY, SETH, etc., derives from a natural hypothesis of analytical simplicity.

Instead, TOLOTTI [5] has constructed a different type of thermo-dynamic potential, starting from a hypothesis of physical simplicity.

In fact, TOLOTTI's idea derives from a remark of SIGNORINI (2), who asserted that HADAMARD's statement, accepted by DUHEM, that the quadric of polarisation for waves of discontinuity is always a real ellipsoid, is not acceptable in finite strain. Thence arises doubt whether in all cases of large deformation the quadric continues to be real, as it always is in small strain. TOLOTTI constructed an example of a quadratic theory failing to yield a real ellipsoid. However, to avoid misunderstanding, it is better to explain the matter.

In fact, HADAMARD proved that stability of equilibrium, according to his definition, suffices that the quadric of polarisation at every point and for every direction of propagation be a real ellipsoid[1]. However, if one assumes, as is natural, that states of unstable equilibrium are possible for an elastic body subject to external forces, and if one assumes also that in unstable cases wave propagation is possible, it is plainly necessary to go beyond HADAMARD's theory [see also TRUESDELL, 9].

Since a natural state of equilibrium is stable (see the *basic property*, page 23), the quadric of polarisation in such a state is a real ellipsoid. For finite strain starting from a natural state, the equilibrium remains stable and the quadric of polarisation remains a real ellipsoid so long as the strain is not too large, that is, so long as a factor to which the external forces are assumed proportional does not become larger than a critical value. For greater values, unstable equilibria exist, and the quadric of polarisation need no longer be a real ellipsoid.

It is appropriate to mention that, according to the classical point of view, in the linear theory of elasticity the above quadric is always a real ellipsoid, and both longitudinal and shear waves are always possible. However, the intimate connection between the kind of equilibrium and the type of the quadric of polarisation should stimulate deeper study of the problem, also because even in the infinitesimal theory states of unstable equilibrium are found by considerations going beyond the linear theory.

In the case of finite strain, the assumption that at least one real speed of propagation must exist at any point and for any direction suggests that the elastic properties of an arbitrary but fixed body may no longer be expressed by a certain kind of elastic potential when the strain is large enough that the quadric of polarisation ceases to be a real quadric.

However, in this field TRUESDELL has made the interesting remark that the inability of the material to transmit a certain kind of waves may be interpreted as the fact that such waves offer a mechanism for carrying a body from a state of equilibrium to another one.

TOLOTTI has studied the problem of finding the most general isothermal potential for which the quadric of polarisation is a real ellipsoid of rotation around the direction of propagation of the wave.

---

[1] See also CATTANEO.

With reference to bodies homogeneous and isotropic in C^*, and considering the direct displacement $C^* \to C$, TOLOTTI has demonstrated that for the above-stated property to hold a necessary and sufficient condition is

$$W_\tau = F_\tau(D^{(\tau)}) + a_\tau I_1^{(\tau)}, \tag{3.80}$$

with

$$a_\tau > 0. \tag{3.81}$$

The theory based on (3.80), (3.81) exhibits the principal properties expressed in § 2 if and only if the following conditions are satisfied:

a) $F_\tau(D^{(\tau)})$ tends toward $+\infty$ if $D^{(\tau)}$ tends to zero; is a decreasing function of $D^{(\tau)}$; is equal to zero for $D^{(\tau)} = 1$; tends to a finite limit when $D^{(\tau)}$ tends toward infinity.

b) $D^{(\tau)} \dfrac{dF_\tau}{dD^{(\tau)}}$ tends toward $-\infty$ if $D^{(\tau)}$ tends to zero; is an increasing function of $D^{(\tau)}$; and tends to zero when $D^{(\tau)}$ tends toward infinity, while $\left[\dfrac{dF_\tau}{dD^{(\tau)}}\right]_{D_{=1}^{(\tau)}} = -a_\tau$.

If $a_\tau = 0$, the above theory reduces to that of perfect fluids. The thermodynamic potential corresponding to (3.80), (3.81) according to (3.25), (3.28) is

$$J^{(D)} = \frac{1}{k_T} F_T\left(\frac{k_T}{k_\tau} D^{(\tau)}\right) + \frac{a_T}{k_T}\left[\frac{l_\tau^2}{l_T^2}\left(\frac{3}{2} + I_1^{(\tau)}\right) - \frac{3}{2}\right] + \varphi(T), \tag{3.82}$$

where $\varphi(T)$ is expressed by (3.37).

§ 7. A type of thermodynamic potential proposed by Bordoni

1. A type of thermodynamic potential resulting from an assumption of analytical simplicity was proposed by P. G. BORDONI [1] in 1953.

Bearing in mind the fact that the Eulerian components of stress cannot be linear functions of the components of the inverse displacement, BORDONI has considered the problem of determining whether relations of the type

$$X_{rs} = [p(\tau, T) + \alpha'(\tau, T) I_1' + 2\beta'(\tau, T) \varepsilon_{rs}'] f(I_1', I_2', D') \tag{3.83}$$

may exist between X_{rs} and ε_{rs}'. In (3.83) $f'(I_1', I_2', D')$ cannot be a constant, since the X_{rs} cannot be linear functions of ε_{pq}'. By (3.14), it is necessary that

$$\begin{rcases} l' = (p + \alpha' I_1') f(I_1', I_2', D'), \\ m' = \beta' f(I_1', I_2', D'), \\ n' = 0, \end{rcases} \tag{3.84}$$

while from (3.15), (3.84) one shows that $J'^{(D')}$ is independent of I_2' and is restricted only by the equation

$$\left(\frac{\alpha'}{\beta'} I_1' + \frac{p}{\beta'} - 1\right) \frac{\partial J'}{\partial I_1'} = D' \frac{\partial J'}{\partial D'}. \tag{3.85}$$

The most general solution of (3.85) is

$$J' = \Psi(y) - q \tag{3.86}$$

where $\Psi(y)$ is an arbitrary function of the parameter

$$y = \frac{\beta'^2}{\alpha' k_\tau} \left[1 - bD'^{-\frac{\alpha'}{\beta'}} \left(1 - \frac{p}{\beta'} - \frac{\alpha'}{\beta'} I_1'\right)\right]. \tag{3.87}$$

Consequently, (3.83) becomes

$$X_{rs} = bD'^{\frac{\alpha'+\beta'}{\beta'}} [p + \alpha' I_1' + 2\beta' \varepsilon_{rs}'] \frac{d\Psi}{dy}. \tag{3.88}$$

Put $\alpha'(\tau, \tau) = \lambda_\tau$, $\beta'(\tau, \tau) = \mu_\tau$, $p(\tau, \tau) = 0$; it is easy to show that in the case of small transformations y becomes the classical isothermal elastic potential

$$y = \frac{1}{2 k_\tau} [(\lambda_\tau + 2\mu_\tau) I_1^2 - 4\mu_\tau I_2] + \cdots \tag{3.89}$$

on the assumption that $b(\tau, \tau) = 1$, $\left[\dfrac{d\Psi}{dy}\right]_{\substack{\varepsilon_{rs}=0 \\ T=\tau}} = 1$,

$$\Psi(y) = y + \cdots. \tag{3.90}$$

The isothermal potential corresponding to (3.86) is

$$W_\tau = k_\tau \Psi(y_\tau) = k_\tau (y_\tau + \cdots) = \frac{\mu_\tau^2}{\lambda_\tau} \left[1 - D'^{-\frac{\lambda_\tau}{\mu_\tau}} \left(1 - \frac{\lambda_\tau}{\mu_\tau} I_1'\right)\right], \tag{3.91}$$

whence it follows that λ_τ, μ_τ must satisfy the conditions (3.52). However, generally, the inequalities

$$\lambda_\tau > 0, \qquad \mu_\tau > 0 \tag{3.92}$$

are sufficient[1] that $\Psi(y_\tau) \geqslant 0$ when Ψ has the same sign as y_τ. The expression of the thermodynamic potential corresponding to (3.86) is

$$J' = \Psi\left\{\frac{\mu_T^2}{\lambda_T} \frac{D'}{k_T} \left(\frac{l_T}{l_\tau}\right)^3 \left[1 - \left(\frac{l_T}{l_\tau}\right)^{\frac{3\lambda_T}{\mu_T}+2} D'^{\frac{\lambda_T}{\mu_T}} \left(1 - \left(1 + \frac{3}{2} \frac{l_T}{l_\tau}\right) \left(\frac{l_\tau^2 + l_\tau^2}{l_T^2}\right) - \right.\right.\right.$$
$$\left.\left.\left. - \frac{\lambda_T}{\mu_T} I_1'\right)\right]\right\} - e \int_{T_0}^{T} d\tau' \int_{\tau_0}^{\tau'} c_p(\tau'') \frac{d\tau''}{\tau''}. \tag{3.93}$$

---

[1] In a later paper BORDONI [4] showed that condition (3.92, 1) is not necessary; then (3.52) are necessary and sufficient conditions that W_τ be positive definite.

It is possible to demonstrate that a necessary and sufficient condition in order for J' to satisfy the Helmholtz postulate is

$$\frac{d^2 \mu_T}{d T^2} < 0 . \tag{3.94}$$

2. Bordoni, supposing that $\Psi(y) = y$, has applied his theory to the isothermal cases of uniform pressure and simple extension. In the first case, if $- E'$ is the common value of the three principal components of strain of the inverse displacement, one finds that

$$T = (3 \lambda + 2 \mu) E' (1 - 2 E')^{\frac{3\lambda + \mu}{2\mu}} , \tag{3.95}$$

where T is the common value of the principal stresses referred to the unit surface of C_τ^*.

From (3.95) one deduces the existence of a critical value for the stress $\overline{T}$. An analogous thing happens in the second case. Further, if $\Psi = y$, the coefficient of lateral contraction coincides with that of the linear theory.

3. The same author, supposing always $\Psi = y$, has applied his theory also to the case of adiabatic finite transformations (Bordoni [2]) and has observed that while in the linear case any transformation without variation of volume $[\varepsilon_\tau^{\cdot r} = 0]$ is isothermal, in the case of adiabatic large deformations this is not so, since invariability of volume does not necessarily imply invariability of temperature.

Another application of the theory was made by the author to the non-isothermal problem of uniform pressure (Bordoni [3]) with results which are valid also if $\Psi(y) \neq y$. Among other things, it is shown that the expression for the isothermal compressibility coincides with that in linear theory and that the Grüneisen coefficient given by the theory agrees well with that found experimentally for those bodies for which the ratio of the Lamé constants is close to 1. For variation of temperature at constant volume the theory gives a pressure agreeing with that deduced from the statistical theory of the solid state.

Chapter IV

Transformations Depending on a Parameter. Successive Equations of Elasticity

§ 1. Displacements depending on a parameter

1. The study of the difficult analytic problems relative to the fundamental set of equations of static elasticity becomes easier if the transformation $C^* \to C$ is supposed to depend upon a parameter.

As will be seen, this assumption makes it possible not only to establish procedures of successive approximation but also to deal with the relevant existence problem [Chap. V]. Some years ago A. SIGNORINI [4], considering displacements depending on a parameter, showed that the calculation of Y_{rs} may be made to depend upon the solution of an infinite number of successive linear systems of differential equations, and he also established uniqueness theorems and indicated some cases of incompatibility. Finally, by considering elastic displacements depending upon a parameter it is possible to give precise and elegant form to the usual methods based on supposing that the elastic isothermal potential is expressible as a power series and then taking its first few terms.

Let us suppose, therefore, that the transformation $C^* \to C$ depends upon the parameter θ. That is, let the displacement u_r and temperature T be functions of θ, and suppose that $u_r \equiv 0$ and $T = T_0$ for $\theta = 0$. Likewise ε_{rs}, Y_{rs}, X_{rs} and E are functions of θ. For any function g depending upon θ put

$$g^{(n)} = \left(\frac{d^n g}{d\theta^n} \right)_{\theta=0}. \tag{4.1}$$

Let the equality

$$\frac{\partial^p g^{(n)}}{\partial y_r^{(p)}} = \left(\frac{\partial^p g}{\partial y_r^p} \right)^{(n)} \tag{4.2}$$

be valid for any derivative which it is necessary to consider. I shall say that the displacement is of order h if u_r, ε_{rs}, Y_{rs}, etc. have derivatives up to and including the hth order at $\theta = 0$. From (1.11) it follows, in particular, that

$$\left. \begin{array}{ll} \varepsilon_{rs}^{(0)} = 0, & \varepsilon_{rs}^{(1)} = \dfrac{1}{2} \left(u_{r,s}^{(1)} + u_{s,r}^{(1)} \right), \\[2mm] \varepsilon_{rs}^{(2)} = \dfrac{1}{2} \left(u_{r,s}^{(2)} + u_{s,r}^{(2)} \right) + u_{i,r}^{(1)} u_{\cdot,s}^{(1) i}. & \end{array} \right\} \tag{4.3}$$

Often I shall use the symbols with one index according to the scheme (1.12) and set $\varepsilon_7 = T - T_0$.

The thermodynamic potential depends upon θ through the displacement, and it is possible to verify that

$$\left(\frac{\partial J}{\partial \varepsilon_j} \right)^{(n)} = \frac{1}{n+1} \frac{\partial J^{(n+1)}}{\partial \varepsilon_j^{(1)}} \quad (j = 1, 2, \cdots, 7; \ n = 0, 1, \cdots, h-1). \tag{4.4}$$

2. From equalities (2.24), using for Y_{rs} the symbols with one index only, namely

$$Y_r = Y_{rr}, \qquad Y_{r+3} = Y_{r+1\,r+2}, \tag{4.5}$$

we may show that for any solution of order $h-1$,

$$Y_j^{(n-1)} = -\frac{1}{n}\frac{\partial J^{(n)}}{\partial \varepsilon_j^{(1)}}, \quad eE^{(n-1)} = -\frac{1}{n}\frac{\partial J^{(n)}}{\partial T^{(1)}}$$

$$(j = 1, 2, \cdots, 6; \quad n = 1, 2, \cdots, h). \quad (4.6)$$

For an elastic body, from condition (2.25) it follows that

$$\int_{C^*} J^{(2s-1)} dC^* = 0, \quad \int_{C^*} J^{(2s)} dC^* > 0 \quad (s = 1, 2\cdots) \quad (4.7)$$

for any isothermal transformation which is not rigid. It is easy to see that

$$J^{(n)} = \sum_l \left(\frac{\partial J}{\partial \varepsilon_l}\right)^{(0)} \varepsilon_l^{(n)} + (n-\delta_{n2})\sum_{l,m}\left(\frac{\partial^2 J}{\partial \varepsilon_l \partial \varepsilon_m}\right)^{(0)} \varepsilon_l^{(n-1)}\varepsilon_m^{(1)} + \cdots. \quad (4.8)$$

One deduces that $Y_j^{(n-1)}$, $E^{(n-1)}$ are derived from the potential $J^{(n)}(\varepsilon^{(1)}\cdots, \varepsilon_j^{(n)}; T)$ according to (4.6) and are linear functions of $\varepsilon_j^{(n-1)}$, $T^{(n-1)}$. In particular,

$$J^{(1)} = \sum_{l=1}^{7}\left(\frac{\partial J}{\partial \varepsilon_l}\right)^{(0)}\varepsilon_l^{(1)},$$

$$\left.\begin{array}{l} \\ \end{array}\right\} \quad (4.9)$$

$$J^{(2)} = \sum_{l=1}^{7}\left(\frac{\partial J}{\partial \varepsilon_l}\right)^{(0)}\varepsilon_l^{(2)} + \sum_{l,m=1}^{7}\left(\frac{\partial^2 J}{\partial \varepsilon_l \partial \varepsilon_m}\right)^{(0)}\varepsilon_l^{(1)}\varepsilon_m^{(1)},$$

and therefore

$$Y_j^{(0)} = -\left(\frac{\partial J}{\partial \varepsilon_j}\right)^{(0)}, \quad eE^{(0)} = -\left(\frac{\partial J}{\partial T}\right)^{(0)}, \quad (4.10)$$

$$Y_j^{(1)} = -\sum_{l=1}^{6}\left(\frac{\partial^2 J}{\partial \varepsilon_l \partial \varepsilon_j}\right)^{(0)}\varepsilon_l^{(1)} - \left(\frac{\partial^2 J}{\partial \varepsilon_j \partial T}\right)^{(0)}(T-T_0),$$

$$\left.\begin{array}{l} \\ \end{array}\right\} \quad (4.11)$$

$$eE^{(0)} = -\sum_{l=1}^{6}\left(\frac{\partial^2 J}{\partial \varepsilon_l \partial T}\right)^{(0)}\varepsilon_l^{(1)} - \left(\frac{\partial^2 J}{\partial T^2}\right)^{(0)}(T-T_0).$$

Equations (4.11, 1) give $Y_{rs}^{(1)}$ as linear expressions in $u_{p,q}^{(1)}$ and, in principle, express Hooke's law for any isothermal transformation. It may be useful to observe that from (4.6, 1), (4.8) follows

$$Y_j^{(n-1)} = -\sum_{l=1}^{6}\left(\frac{\partial^2 J}{\partial \varepsilon_l \partial \varepsilon_j}\right)^{(0)}\varepsilon_l^{(n-1)} \text{ plus terms which do not depend upon } \varepsilon_l^{(n-1)}, \quad (4.12)$$

by which one recognizes that the coefficients of the part of $Y_j^{(n-1)}$ which contains $\varepsilon_l^{(n-1)}$ are independent of n. In particular, for isotropic bodies, Lamé's coefficients of classical elasticity always occur in that part, for any n.

§ 2. Successive systems. Linear Elasticity

1. In the mechanics of continuous media it is fundamental to assume the existence of a state of *free equilibrium* subject to no external forces and at uniform temperature, for any temperature belonging to a certain interval. Even if the idea of a body in equilibrium subject to no external forces is scarcely acceptable in engineering applications, one may take such an assumption as expressing the first property of the thermodynamic potential $J(\varepsilon, T)$, at least for bodies undergoing reversible transformations: J must be such that equations (2.12), (2.13), (2.24) admit the solution $u_r \equiv 0$ if external forces are absent. Let us, then, assume the existence of a state of free equilibrium and identify C^* with it.

Moreover, I suppose that the magnitudes of the external forces are proportional to the parameter θ, thus writing θF_r and θf_r^* instead of F_r, f_r^*. Therefore the equations of equilibrium are

$$(Y_{is}\, x_r^{\cdot i})^{\cdot s} = k^* \theta F_r, \tag{4.13}$$

$$Y_{is}\, x_r^{\cdot i} N^s = \theta f_r^*. \tag{4.14}$$

Any solution of (2.24), (4.13), (4.14) is a function of θ. I assume that these equations have solutions of order $h \geqslant 1$ for any θ which satisfies the inequality

$$0 \leqslant \theta \leqslant \bar{\theta}. \tag{4.15}$$

From (4.13), (4.14) follow the successive systems of equations

$$Y_{rs}^{(0),s} = 0 \quad (\text{in } C^*), \qquad Y_{rs}^{(0)} N_*^s = 0 \quad (\text{on } \Sigma^*); \tag{4.16}$$

$$\left.\begin{array}{ll} Y_{rs}^{(1),s} + (Y_{lm}^{(0)} u_r^{(1),m})^{\cdot l} = k^* F_r & (\text{in } C^*), \\ Y_{rs}^{(1)} N_*^s + Y_{lm}^{(0)} u_r^{(1),m} N_*^l = f_r^* & (\text{on } \Sigma^*); \end{array}\right\} \tag{4.17}$$

$$\cdots\cdots\cdots\cdots\cdots\cdots$$

$$\left.\begin{array}{ll} Y_{rs}^{(n),s} + (Y_{lm}^{(0)} u_r^{(n),m})^{\cdot l} = k^* F_r^{(n)} & (\text{in } C^*) \\ Y_{rs}^{(n)} N_*^s + Y_{lm}^{(0)} u_r^{(n),m} N_*^l = f_r^{*(n)} & (\text{on } \Sigma^*) \end{array}\right\} \quad (n = 2, 3, \ldots, h), \tag{4.18}$$

where

$$\left.\begin{array}{l} k^* F_r^{(n)} = -\sum\limits_{q=1}^{n-1} \binom{n}{q} (Y_{lm}^{(q)} u_r^{(n-q),m})^{\cdot l} \\[2mm] f_r^{(n)} = -\sum\limits_{q=1}^{n-1} \binom{n}{q} Y_{lm}^{(q)} u_r^{(n-q),m} N_*^l. \end{array}\right\} \quad (n = 2, 3, \cdots, h), \tag{4.19}$$

It is evident that the $Y_{rs}^{(0)}$ coincide with the values of the X_{rs} in C^*. These values are all equal to zero if the body is homogeneous or isotropic in C^*.

Along with (4.16), (4.17), (4.18), (4.19) we have also the relations of symmetry

$$Y_{rs}^{(n)} = Y_{sr}^{(n)} \qquad (n = 0, 1, \cdots, h). \tag{4.20}$$

(4.10), (4.16) are conditions on the thermodynamic potential: it must satisfy the equations

$$
\left.
\begin{aligned}
\sum_s \frac{\partial}{\partial y_s} \left(\frac{\partial J}{\partial \varepsilon_{rs}}\right)^{(0)} &= 0 \qquad \text{(in } C^*) , \\
\sum_s \left(\frac{\partial J}{\partial \varepsilon_{rs}}\right)^{(0)} N^s_* &= 0 \qquad \text{(on } \Sigma^*) .
\end{aligned}
\right\} \tag{4.21}
$$

2. (4.11), (4.17) are the linear equations of the classical theory of elasticity, where the stress is derived from a potential which is a quadratic function (4.9, 2) satisfying (4.7, 2) when $s = 1$.

Expressing u_r by MacLaurin's formula, we have

$$
u_r = \sum_{i=1}^{h-1} u_r^{(i)} \frac{\theta^i}{i!} + R_h . \tag{4.22}
$$

It is clear that the classical linear theory gives the solution of the fundamental problem of static elasticity if, and only if, the two following circumstances hold: a) R_2 may be considered negligible, and thus u_r may be expressed as $u_r^{(1)} \theta$; b) non-linear terms in $u_r^{(1)} \theta$ are supposed negligible in the expression of ε_{rs} and Y_{rs} in such a way that according to (4.6, 1) $Y_{rs}^{(j)}$ are to be considered zero for $j \geqslant 2$. Instead, if one supposes R_p negligible for $2 < p \leqslant h$, the transformation $C^* \to C$ is represented by equations (4.17), (4.18), each of which is linear in $u_r^{(n)}$ ($n = 1, 2 ..., p - 1$), and its known terms depend on the solutions of the previous sets of equations, provided products $\theta^\varrho u_r^{(1) \mu_1} u_s^{(2) \mu_2} \cdots u_t^{(p-1) \mu_{p-1}}$, where $\varrho = \mu_1 + 2\mu_2 + \cdots + (p-1)\mu_{p-1} \geqslant p$, be negligible.

In this instance there are $p - 1$ sets of equations (4.17), (4.18), and knowledge of the functions $J^{(1)}, J^{(2)}, \cdots J^{(p)}$ is necessary to determine the solution. It may happen that $h = \infty$, and the solution of the fundamental problem of static elasticity is expressible as a power series in θ [Chap. V].

§ 3. Necessary conditions for the existence of solutions of the successive systems of equations

1. Plainly, from (4.17) follows

$$
\left.
\begin{aligned}
\int_{C^*} k^* F_r \, dC^* + \int_{\Sigma^*} f_r^* \, d\Sigma^* &= 0 , \\
\int_{C^*} k^* (F_r y_t - F_t y_r) \, dC^* + \int_{\Sigma^*} (f_r^* y_t - f_t^* y_r) \, d\Sigma^* &= 0 .
\end{aligned}
\right\} \tag{4.23}
$$

Equations (4.23) mean that the external forces characterized by the vectors $k^* \boldsymbol{F}, \boldsymbol{f}^*$ are in equilibrium if those vectors are applied to the points of C^*. That is possible for at least four special orientations of C^*, according to a theorem of DA SILVA.

From (4.16), (4.18) follow the necessary conditions

$$\int_{C^*} k^* F_r^{(n)} dC^* + \int_{\Sigma^*} f_r^{*\,(n)} d\Sigma^* = 0 , \qquad (4.24)$$

$$\int_{C^*} k^* (F_r^{(n)} y_t - F_t^{(n)} y_r) dC^* +$$

$$+ \int_{\Sigma^*} (f_r^{(n)} y_t - f_t^{(n)} y_r) d\Sigma^* = 0 \quad (n = 2, 3, \cdots, h) . \qquad (4.25)$$

According to (4.19), and by the symmetry of $Y_{rs}^{(n)}$, one sees that (4.25) may be expressed in the form

$$\sum_{q=1}^{n-1} \binom{n}{q} \int_{C^*} [Y_{rm}^{(q)} u_t^{(n-q),\,m} - Y_{tm}^{(q)} u_r^{(n-q),\,m}] dC^* = 0 , \quad (n = 2, 3, \cdots, h) . \qquad (4.26)$$

Eqs. (4.26), in view of (4.17), after development become

$$\int_{C^*} [u_i^{(n-1)} F_r - u_r^{(n-1)} F_i] k^* dC^* +$$

$$+ \int_{\Sigma^*} [u_i^{(n-1)} f_r^* - u_r^{(n-1)} f_i^*] d\Sigma^* = 0 \quad (n = 2, 3, \cdots, h) . \qquad (4.27)$$

which may be put into the vectorial form

$$\int_{C^*} \mathbf{s}^{(n-1)} \times k^* \boldsymbol{F} dC^* + \int_{\Sigma^*} \mathbf{s}^{(n-1)} \times \boldsymbol{f}^* d\Sigma^* = 0 \quad (n = 2, 3, \cdots, h) , \qquad (4.28)$$

where $\mathbf{s}^{(n-1)}$ is the vector whose components are $u_r^{(n-1)}$.

While it is possible that (4.23) are satisfied at least for some special orientation of C^*, (4.28) represent real necessary conditions for the existence of solutions of the sets of equations (4.18). In fact, bearing in mind that the external forces are in equilibrium in C^*, we see that (4.28) state that they are in equilibrium also in C.

2. In general it is possible to satisfy (4.27), as I shall demonstrate at once, but there are special cases where it is not so. If (4.27) are compatible only when $n \leq \bar{n}$, then the solutions of the basic equations (4.13), (4.14) [bearing in mind (2.24)] are at most of order $\bar{n}$, and it is impossible to express u_r by a power series in the parameter θ, since a necessary condition is $h = \infty$, $\bar{n} = \infty$.

Let $\mathbf{s}^{(1)}$, $\mathbf{s}^{(2)}$, $\cdots$, $\mathbf{s}^{(h)}$ be h vectors which are solutions of equations (4.6), (4.16), (4.17), (4.18) [for example, satisfying the conditions $u_r^{(q)} (0, 0, 0) = 0$, $u_{r,\,r+1}^{(q)} - u_{r+1,\,r}^{(q)} = 0$]. It is evident that the vectors $\mathbf{s}^{(q)} + \boldsymbol{\omega}^{(q)} \times OP^*$ are also solutions of the same equations, if the vectors $\boldsymbol{\omega}^{(q)}$ do not depend on y_r. The meaning of (4.27) is: *a set of vectors $\boldsymbol{\omega}^{(q)}$ $(q = 1, 2, \cdots, h)$ satisfying them must exist in order that equations (4.18) be compatible.*

Let

$$a_{rs} = -\frac{1}{C^*}[\int_{C^*} k^* F_r \, y_s \, dC^* + \int_{\Sigma^*} f_r^* \, y_s \, d\Sigma^*]\qquad(4.29)$$

be the astatic coordinates of the load, and set $\sigma = a_{11} + a_{22} + a_{33}$ (SIGNORINI [1]).

By (4.23, 2) it follows that $a_{rs} = a_{sr}$, and (4.27) becomes

$$(a^i_{.r} - \delta^i_{.r}\sigma)\,\omega_i^{(s-1)} = \frac{1}{C^*}[\int_{C^*} (u_{r+2}^{(s-1)} F_{r+1} - u_{r+1}^{(s-1)} F_{r+2})\,k^* dC^* +$$

$$+ \int_{\Sigma^*} (u_{r+2}^{(s-1)} f_{r+1}^* - u_{r+1}^{(s-1)} f_{r+2}^*)]\,d\Sigma^* \qquad (s = 2, 3, \cdots, h)\,. \qquad(4.30)$$

If

$$A = \|a_{rs} - \delta_{rs}\sigma\| \neq 0\,, \qquad(4.31)$$

(4.30) are compatible, and the vectors $\omega^{(s-1)}$ are all determinated by them. Instead, if $A = 0$, (4.30) are conditions for the vectors $s^{(1)}, \ldots, s^{(h-1)}$, conditions which may or may not be satisfied.

3. In particular, (4.30) determine the vector $\omega^{(1)}$ if $A \neq 0$. This means that an arbitrary rigid rotation [if it is permitted by the constraints] is not allowed in the solution of the equations of linear elasticity, as commonly thought, but rather the rotation is determined by (4.30) for $s = 2$ (SIGNORINI [4]).

If the solution of the basic equations is of order h, the vectors $s^{(1)}, \ldots,$ $s^{(h-1)}$ are determined, while the vector $s^{(h)}$ is determined only to within an infinitesimal rotation. In the case when $A = 0$ but (4.30) are compatible, the vectors $\omega^{(s)}$ are partially or totally indeterminate.

The problem of the compatibility of the succesive systems of equations has been studied by TOLOTTI [2], who has introduced the notion of *principal orientation of order n*. In fact when $A = 0$ the external forces applied to the points of C^* admit an axis of equilibrium, that is, an axis a such that any rotation of C^* around a does not destroy the equilibrium of the external forces if the directions and magnitudes of the vectors $k^* F, f^*$ are not changed. The orientations of C^* for which (4.30) are compatible for $s = 2, 3, \ldots, n$, but not for $s = n + 1$, are the principal orientations of order $n - 1$. Therefore, the condition that the orientation of C^* is principal of order $n - 1$ is necessary in order that the solution of the basic equations of static elasticity be of order n. If the external forces satisfy the equations of equilibrium (4.23), then C^* is a principal orientation of order zero. If $A = 0$ and (4.30) are not compatible for a given orientation of C^*, it may happen that (4.30) become compatible when the orientation of C^* is changed. This difficult question has been studied by

TOLOTTI [1], who has considered the problem of characterizing the principal orientations when there are axes of equilibrium. TOLOTTI has considered in greater detail the case of the first order and has demonstrated that a principal orientation of the first order always exists, except possibly in the case when every straight line is an axis of equilibrium [$a_{rs} = 0$]. In other words, if (4.30) are not compatible for the given orientation C^* for $s = 2$, but not all a_{rs} are zero, it is possible to find a special orientation of C^* for which (4.30) become compatible for $s = 2$. If, however, all a_{rs} are zero, there is not necessarily any orientation such that (4.30) are compatible also for $s = 2$ (see the observations of TOLOTTI [1] on an example of SIGNORINI). In this case, the real meaning of the classical linear theory of elasticity is doubtful, at least if one thinks that u_r ought to have derivatives with respect to θ up to and including the second order.

If $A \neq 0$, axes of equilibrium are absent for the external forces applied to C^*, and all orientations of C^* are principal of infinite order. In this case (4.30) are compatible for any value of s and h, and one may presume that u_r is expressible as a power series in θ. On this hypothesis SIGNORINI ([4], p. 11) has demonstrated a uniqueness theorem by considering the successive sets of equations (4.16), (4.17), (4.18).

§ 4. On the conditions of compatibility of the successive systems of equations when constraints are present

1. What was said in the last section refers to the case when the external forces are everywhere known. It is interesting to make some remarks if constraints are present on a portion of the boundary. I suppose that the constrained portion Σ_1^* of Σ^* is known. Then suppose that the external forces are known only on the portion $\Sigma_2^* = \Sigma^* - \Sigma_1^*$, remaining unknown on Σ_1^*. On this hypothesis equations (4.13), 4.14) are to be replaced by

$$(Y_{is}\, x_r^{\cdot i})^{\cdot s} = k^* \theta F_r \qquad \text{(in } C^*), \qquad (4.32)$$

$$Y_{is}\, x_r^{\cdot i}\, N_*^s = \begin{cases} \varPhi_r^* & \text{(on } \Sigma_1^*), \\ \theta f_r^* & \text{(on } \Sigma_2^*), \end{cases} \qquad (4.33)$$

where $\varPhi^* d\Sigma_1^*$ denotes the unknown force exerted by the constraint on $d\Sigma_1^*$. Plainly the corresponding force at the surface element $d\Sigma_1^*$ of C is $\varPhi d\Sigma_1$, where a relation of the kind (2.14) subsists between $\varPhi$ and $\varPhi^*$. The vector $\varPhi^*$ is a function of θ which vanishes when $\theta = 0$.

Let us suppose that $\varPhi^*$ is differentiable for $\theta = 0$. Further, to simplify matters, let us suppose that C^* is a state of natural equilibrium [$Y_{rs}^{(0)} \equiv 0$]. While the first equations of the successive systems (4.17), (4.18) are still valid, it is necessary to introduce $\boldsymbol{\Phi}^*$ into the second of them. Specifically these last equations are to be changed into

$$Y_{rs}^{(n)} N_*^s = \begin{cases} f_r^{(n)} + \Phi_r^{(n)} & \text{(on } \Sigma_1^*) \\ f_r^{(n)} & \text{(on } \Sigma_2^*) \end{cases} \quad (n = 1, 2, \ldots, h), \qquad (4.34)$$

where $f_r^{(1)} = f_r^*$. By a procedure analogous to that used in deducing (4.28), one now finds that the equation

$$\int_{C^*} s^{(n-1)} \times k^* F \, dC^* + \int_{\Sigma_2^*} s^{(n-1)} \times f^* \, d\Sigma_2^* + \int_{\Sigma_1^*} s^{(n-1)} \times \Phi^{(1)} d\Sigma_1^* +$$

$$+ \frac{1}{n} \int_{\Sigma_1^*} OP^* \times \Phi^{(n)} d\Sigma_1^* = 0 \qquad (4.35)$$

is a necessary condition for the integrability of the successive differential system of index n.

Formally, (4.35) does not coincide with the condition of equilibrium of moments for the body in its actual state, as is the case for (4.28). In fact, it is possible to recognize that the condition of equilibrium of moments for C is equivalent to the equation

$$n \int_{C^*} s^{(n-1)} \times k^* F \, dC^* + n \int_{\Sigma_2^*} s^{(n-1)} \times f^* \, d\Sigma^* + \int_{\Sigma_1^*} \sum_{p=1}^{n-1} \binom{n}{p} s^{(n-p)} \times \Phi^{(p)} d\Sigma_1^* +$$

$$+ \int_{\Sigma_1^*} OP^* \times \Phi^{(n)} d\Sigma_1^* = 0 \qquad (n = 2, 3, \ldots, h), \quad (4.36)$$

but one may demonstrate that

$$\sum_{p=2}^{n-1} \binom{n}{p} \int_{\Sigma_1^*} s^{(n-p)} \times \Phi^{(p)} d\Sigma_1^* = 0 \qquad (n = 3, 4, \ldots, h) . \qquad (4.37)$$

Thus (4.35) and (4.36) are equivalent.

To justify (4.37) it is useful to observe that from (4.17, 1), (4.18, 1), and (4.34) follows

$$\int_{\Sigma_1^*} \sum_{p=2}^{n-1} \binom{n}{p} [\Phi_r^{(p)} u_i^{(n-p)} - \Phi_i^{(p)} u_r^{(n-p)}] \, d\Sigma_1^* = \sum_{p=2}^{n-1} \binom{n}{p} \int_{\Sigma^*} [Y_{rs}^{(p)} u_i^{(n-p)} -$$

$$- Y_{is}^{(p)} u_r^{(n-p)}] N_*^s d\Sigma^* - \sum_{p=2}^{n-1} \binom{n}{p} \int_{\Sigma_2^*} [f_r^{(p)} u_i^{(n-p)} - f_i^{(p)} u_r^{(n-p)}] \, d\Sigma^* +$$

$$+ \sum_{p=2}^{n-1} \binom{n}{p} \int_{C^*} [Y_{rs}^{(p), s} u_i^{(n-p)} - Y_{is}^{(p), s} u_r^{(n-p)}] \, dC^* -$$

$$- \sum_{p=2}^{n-1} \binom{n}{p} \int_{C^*} k^* [F_r^{(p)} u_i^{(n-p)} - F_i^{(p)} u_r^{(n-p)}] \, dC^* . \qquad (4.38)$$

It is possible to demonstrate that the second member of (4.38) is equal to zero.

2. Along with the conditions of integrability (4.35) one has to consider the others, namely,

$$\int_{\Sigma_1^*} \Phi_r^{(n)} d\Sigma_1^* = 0 \qquad\qquad (n = 2,3,\ldots,h) \qquad\qquad (4.39)$$

which follow at once from the successive systems (4.18, 1), (4.34) or, if one prefers, from the first fundamental equation of statics. Naturally, with (4.35), (4.39) are to be associated the conditions of integrability of the first of the successive systems:

$$\int_{C^*} k^* \, F \, dC^* + \int_{\Sigma_2^*} f^* \, d\Sigma_2^* + \int_{\Sigma_1^*} \Phi^{(1)} d\Sigma_1^* = 0 \,,$$

$$\int_{C^*} OP^* \times k^* \, F \, dC^* + \int_{\Sigma_2^*} OP^* \times f^* \, d\Sigma_1^* + \int_{\Sigma_1^*} OP^* \times \Phi^{(1)} d\Sigma_1^* = 0 \,. \qquad (4.40)$$

If the constraint on Σ_1^* is such that the displacement is assigned for the points of Σ_1^* and, analogously to what is done for the forces, is expressed by the vector $\theta \, s'$, where s' is independent of θ, the third integral of (4.35) is equal to zero for any value $n > 2$. In particular, s' is zero, and the third term in (4.35) is absent if the body is clamped at all points of Σ_1^*. Generally, if constraints are present, it is impossible to have cases of incompatibility of the kind considered in the last section, since condition (4.35) is an analytic consequence of the set of equations of index n and depends not only on the index $n - 1$ but also on the index n. Finally, it remains to make sure that the constraints may correspond to forces whose resultant moment of index n satisfies (4.35). For example, if the body is supported on a rigid part of the plane $y_3 = 0$ and the constraint is frictionless and bilateral[1], the following equations are valid on Σ_1^*:

$$u_3^{(n)} = 0 \,, \qquad \Phi_1^{(n)} = \Phi_2^{(n)} = 0 \qquad (n = 1,2,\ldots,h) \,. \qquad (4.41)$$

From (4.35), (4.41) follows

$$\int_{C^*} [u_1^{(n-1)} F_2 - u_2^{(n-1)} F_1] \, k^* \, dC^* + \int_{\Sigma_2^*} [u_1^{(n-1)} f_2^* - u_2^{(n-1)} f_1^*] \, d\Sigma_2^* = 0 \,, \qquad (4.42)$$

representing a real condition on $s^{(n-1)}$, analogous to (4.27), for the integration of the system of equations of index n. Considerations analogous to those in the last section show that (4.42) either determines the rigid rotation of the body around an axis orthogonal to the plane $y_3 = 0$ or — exceptionally — may give rise to a real case of incompatibility. Specifically, let $u_r^{(n-1)}$ be an assigned solution of the system of equations of index $n - 1$ and let $\omega^{(n-1)}$ be a vector perpendicular to plane $y_3 = 0$ and independent

---

[1] In calling a constraint *bilateral* I mean that only equalities are necessary in order to express it analytically, as in the case of the clamping constraint.

of y_i. Plainly the vector having components $[u_1^{(n-1)} - y_2 \omega^{(n-1)}, u_2^{(n-1)} + y_1 \omega^{(n-1)}, u_3]$ satisfies the same set of equations, and (4.42) then becomes

$$\omega^{(n-1)} \left\{ \int_{C^*} (y_1 F_1 + y_2 F_2) \, k^* d C^* + \int_{\Sigma_1^*} (y_1 f_1^* + y_2 f_2^*) \, d \Sigma_2^* \right\} =$$

$$= \int_{C^*} [u_1^{(n-1)} F_2 - u_2^{(n-1)} F_1] \, k^* d C^* + \int_{\Sigma_1^*} [u_1^{(n-1)} f_2^* - u_2^{(n-1)} f_1^*] \, d \Sigma_1^* . \tag{4.43}$$

If $a_{11} + a_{22} \neq 0$ [see (4.29)], (4.43) determines $\omega^{(n-1)}$; if however $a_{11} + a_{22} = 0$, (4.43) represents a real condition on $s^{(n-1)}$, and incompatibility may result.

§ 5. An example of a solution of order one[1]

By way of an example I shall indicate a case where the condition (4.30) is not satisfied even for $s = 2$. Let the elastic body be a cylindrical solid, homogeneous and isotropic, whose height is $2\,l$ and whose lateral surface I denote by σ_l. Let us suppose that the frame of reference is the rectangular right-handed trihedral $O\,y_1\,y_2\,y_3$ with origin at the center of mass of the body and with axes coinciding with those of the ellipsoid of inertia at O, y_3 being taken parallel to the generators.

Let us suppose body forces absent, and as the applied surface force f^* on σ_l let us take the pressure $p_0 y_3 N^*$, while that on the bases $y_3 = \pm l$ is assumed to have the components

$$f_1^* = \pm a y_2, \quad f_2^* = \pm b y_1, \quad f_3^* = 0, \tag{4.44}$$

p_0, a, b being constant. That is, one has a torsional load on the two bases and a non-uniform pressure on the lateral surface. I suppose that the cross-section of the cylinder is an ellipse of semiaxes α, β and that the constants a, b satisfy the condition

$$\frac{a}{\alpha^2} + \frac{b}{\beta^2} = 0. \tag{4.45}$$

According to (4.11) for $T = T_0$, which express Hooke's Law for infinitesimal isothermal transformations, it is easy to see that equations (4.11), (4.17) have the solution

$$\left. \begin{aligned} u_1^{(1)} &= \frac{1}{2\mu(3\lambda + 2\mu)} [(3\lambda + 2\mu)(a - b)\, y_2 - p_0 (\lambda + 2\mu)\, y_1]\, y_3, \\ u_2^{(1)} &= \frac{1}{2\mu(3\lambda + 2\mu)} [(3\lambda + 2\mu)(b - a)\, y_1 - p_0 (\lambda + 2\mu)\, y_2]\, y_3, \\ u_3^{(1)} &= \frac{1}{2\mu(3\lambda + 2\mu)} \Big\{ (3\lambda + 2\mu)(a + b)\, y_1 y_2 + \frac{p_0}{2} [(\lambda + 2\mu) \cdot \\ &\qquad \cdot (y_1^2 + y_2^2) + 2\lambda y_3^2] \Big\}, \end{aligned} \right\} \tag{4.46}$$

---

[1] For other examples see SIGNORINI [5].

where λ, μ are the Lamé constants. One easily verifies that the a_{pq} are equal to zero [see (4.29)]. Therefore, the left-hand member of (4.30) is zero for any s. Instead, for $r = 3$ and $s = 2$ the right-hand member, in view of (4.45), becomes

$$A_3 = -\frac{p_0\,a}{\mu\,(3\lambda + 2\mu)}\left[\frac{l^2}{3}\,(3\lambda + 2\mu)\left(1 + \frac{\beta^2}{\alpha^2}\right) - \frac{\lambda + 2\mu}{4}\,\beta^2\right]. \qquad (4.47)$$

Then, if one excludes an exceptional case, it is impossible to satisfy (4.30) for $s = 2$. One may observe that if only the torsional load is present $[p_0 = 0]$, or if only pressure $[a = b = 0]$, then the right-hand member is zero also for $r = 3$ and $s = 2$. In these cases compatibility holds, and the solution of the fundamental set of equations is of order 2 at least. In fact, this case illustrates what is generally to be expected on account of the non-linearity of condition (4.30) with respect to the external forces [since $u_r^{(s-1)}$ depends upon the external forces]: by superposition of two sets of external forces for each of which $a_{pq} = 0$ and (4.30) are compatible, a new set is obtained for which is $a_{pq} = 0$ but (4.30) are generally incompatible.

§ 6. Successive systems of equations for X_{rs}

1. In previous sections it has been seen that the solution of the fundamental problem of static elasticity, approximately and except in special cases, may be reduced to the integration of certain systems of differential equations analogous to that of the classical linear theory. The difficult analytic questions related to this method will be examined in Chap. V, but still it can be observed that it is possible to obtain approximately also the Eulerian components of stress, X_{rs}. This follows naturally from (2.10), from which results

$$X_{rs} = \theta\,X_{rs}^{(1)} + \frac{\theta^2}{2}\,X_{rs}^{(2)} + \cdots \qquad (4.48)$$

with

$$X_{rs}^{(1)} = -u_p^{(1),\,p}\,Y_{rs}^{(0)} + (u_r^{(1),\,l}\,Y_{ls}^{(0)} + u_s^{(1),\,l}\,Y_{lr}^{(0)}) + Y_{rs}^{(1)},$$

$$X_{rs}^{(2)} = Y_{rs}^{(2)} + 2\,(u_r^{(1),\,l}\,Y_{ls}^{(1)} + u_s^{(1),\,l}\,Y_{lr}^{(1)} - u_l^{(1),\,l}\,Y_{rs}^{(1)}) + \qquad (4.49)$$

$$+\,[2\,(u_p^{(1),\,p})^2 - D^{(2)}]\,Y_{rs}^{(0)} + [(u_r^{(2),\,l} - 2\,u_r^{(1),\,l}\,u_p^{(1),\,p})\,Y_{ls}^{(0)} +$$

$$+\,(u_s^{(2),\,l} - 2\,u_s^{(1),\,l}\,u_p^{(1),\,p})\,Y_{rl}^{(0)}] + 2\,u_r^{(1),\,l}\,u_s^{(1),\,m}\,Y_{lm}^{(0)},$$

etc. In particular, (4.45, 1) are useful to study infinitesimal deformations from an equilibrium state which is stressed, $Y_{rs}^{(1)}$ being given, in this case, by expressions of the kind

$$Y_{rs}^{(1)} = \frac{1}{2} B_{rsih} \left(u^{(1)\,i,h} + u^{(1)\,h,i} \right) . \tag{4.50}$$

However, it may be convenient to derive $X_{rs}^{(1)}$, $X_{rs}^{(2)}$, etc., directly by solutions of the successive systems of equations expressed in terms of them, bearing in mind the relations between X_{rs} and the thermodynamic potential. These differential systems may be obtained by calculating the derivatives of (2.11) with respect to θ for $\theta = 0$ and then substituting them in the successive systems of equations (4.17), (4.18). But it is easier to reach this goal by using the basic equations of static elasticity in the form of Boussinesq [2], bearing in mind (4.48). Supposing for simplicity that the external forces are known, Boussinesq's equations are

$$C_{hs}\, X_r^{\cdot h,\,s} = \theta k^* F_r \qquad \text{(in } C^*\text{)} ,$$
$$C_{hs}\, X_r^{\cdot h}\, N_*^s = \theta f_r^* \qquad \text{(on } \Sigma^*\text{)} . \tag{4.51}$$

Putting $\varDelta = u_{\cdot p}^{\,p}$, from

$$C_{hs} = \delta_{hs}\,(1 + \varDelta) - u_{s,h} + u_{h+1,\,s+1}\,u_{h+2,\,s+2} - u_{h+1,\,s+2}\,u_{h+2,\,s+1} , \tag{4.52}$$

in particular, we see that

$$C_{hs}^{(1)} = \delta_{hs}\,u_p^{(1),\,p} - u_{s,h}^{(1)} . \tag{4.53}$$

On the supposition that C^* is a natural state $[X_{rs}^{(0)} = 0]$, from (4.51), (4.53) the following successive systems are deduced:

$$X_{rs}^{(1),\,s} = k^* F_r , \quad \text{(in } C^*\text{)}; \quad X_{rs}^{(1)}\, N_*^s = f_r^* \qquad \text{(on } \Sigma^*\text{)} , \tag{4.54}$$

$$\left. \begin{aligned} X_{rs}^{(2),\,s} &= 2\,[u_s^{(1),\,h} - u_p^{(1),\,p}\,\delta_s^h]\,X_{rh}^{(1),\,s} && \text{(in } C^*\text{)} , \\ X_{rs}^{(2)}\, N_*^s &= 2\,[u_s^{(1),\,h} - u_p^{(1),\,p}\,\delta_s^h]\,X_{rh}^{(1)}\, N_*^s && \text{(on } \Sigma^*\text{)} , \end{aligned} \right\} \tag{4.55}$$

$$\cdots \cdots \cdots \cdots \cdots \cdots \cdots \cdots$$
$$\cdots \cdots \cdots \cdots \cdots \cdots \cdots \cdots$$

$$X_{rs}^{(n),\,s} = -\sum_{l=1}^{n-1} \binom{n}{l}\, C_{hs}^{(l)}\, X_r^{(n-l)h,\,s} \qquad \text{(in } C^*\text{)} ,$$

$$X_{rs}^{(n)}\, N_*^s = -\sum_{l=1}^{n-1} \binom{n}{l}\, C_{hs}^{(l)}\, X_r^{(n-l)h}\, N_*^s \qquad \text{(on } \Sigma^*\text{)} . \tag{4.56}$$

Naturally, conditions (4.27) are still necessary for the integrability of equations (4.54), (4.55), (4.56), as it is possible to demonstrate, and all the considerations of § 3 about the axes of equilibrium and the principal orientations remain valid also with reference to the calculation of $X_{rs}^{(n)}$.

2. The relations connecting $X_{rs}^{(i)}$ to $u_{pq}^{(l)}$ are to be referred to equations (4.54), (4.55), (4.56). They are obtained according to (3.7), 3.8), (3.9). If C^* is an unstressed state, one has, plainly,

$$D^{(0)} = 1, \quad I_1^{(0)} = 0, \quad D^{(1)} = I_1^{(1)} = u_p^{(1),p}, \quad I_2^{(1)} = 0; \quad (4.57)$$

$$
\begin{aligned}
l_0^{(0)} &= \left(\frac{\partial J^{(D)}}{\partial D}\right)^{(0)} + \left(\frac{\partial J^{(D)}}{\partial I_1}\right)^{(0)} = 0, \\
m^{(0)} &= \left(\frac{\partial J^{(D)}}{\partial I_1}\right)^{(0)} - \frac{1}{2}\left(\frac{\partial J^{(D)}}{\partial I_1}\right)^{(0)}, \\
n^{(0)} &= -2\left(\frac{\partial J^{(D)}}{\partial I_2}\right)^{(0)};
\end{aligned}
\right\} \quad (4.58)
$$

$$
\begin{aligned}
l^{(1)} &= \left(\frac{\partial J^{(D)}}{\partial D}\right)^{(1)} + \left[\left(\frac{\partial J^{(D)}}{\partial D}\right)^{(0)} + \left(\frac{\partial J^{(D)}}{\partial I_2}\right)^{(0)}\right] I_1^{(1)} + \left(\frac{\partial J^{(D)}}{\partial I_1}\right)^{(1)}, \\
m^{(1)} &= \left(\frac{\partial J^{(D)}}{\partial I_1}\right)^{(1)} - \frac{1}{2}\left(\frac{\partial J^{(D)}}{\partial I_2}\right)^{(1)} + \left(\frac{\partial J^{(D)}}{\partial I_2}\right)^{(0)} I_1^{(1)}, \\
n^{(1)} &= -2\left(\frac{\partial J^{(D)}}{\partial I_2}\right)^{(1)};
\end{aligned}
\right\} \quad (4.59)
$$

$$
\begin{aligned}
l^{(2)} = D^{(2)}\left(\frac{\partial J^{(D)}}{\partial D}\right)^{(0)} + 2 I_1^{(1)}\left[\left(\frac{\partial J^{(D)}}{\partial D}\right)^{(1)} + \left(\frac{\partial J^{(D)}}{\partial I_2}\right)^{(1)}\right] + \\
+ \left(\frac{\partial J^{(D)}}{\partial D}\right)^{(2)} + \left(\frac{\partial J^{(D)}}{\partial I_2}\right)^{(0)} I_1^{(2)} + \left(\frac{\partial J^{(D)}}{\partial I_1}\right)^{(2)}, \quad (4.60)
\end{aligned}
$$

etc. It is plain how $\left(\frac{\partial J^{(D)}}{\partial I_1}\right)^{(1)}$ etc., are to be evaluated.

In particular, from (3.8) it follows that

$$
\begin{aligned}
X_{rs}^{(1)} &= -l^{(1)} \delta_{rs} - 2 m^{(0)} v_{rs}^{(1)}, \\
X_{rs}^{(2)} &= [2 l^{(1)} I_1^{(1)} - l^{(2)}] \delta_{rs} + 4 (m^{(0)} I_1^{(1)} - m^{(1)}) v_{rs}^{(1)} - \quad (4.61) \\
&\quad - 2 m^{(0)} v_{rs}^{(2)} - 2 n^{(0)} v_{rt}^{(1)} v_{.s}^{(1)t},
\end{aligned}
$$

where

$$v_{rs}^{(1)} = \frac{1}{2}\left(u_{r,s}^{(1)} + u_{s,r}^{(1)}\right), \qquad v_{rs}^{(2)} = \frac{1}{2}\left(u_{r,s}^{(2)} + u_{s,r}^{(2)} + 2 u_{r,l}^{(1)} u_s^{(1),l}\right). \quad (4.62)$$

Further

$$
\begin{aligned}
I_1^{(2)} &= u_p^{(2),p} + u_{p,q}^{(1)} u^{(1)p,q}, \\
D^{(2)} &= u_p^{(2),p} + (u_p^{(1),p})^2 - u_{p,q}^{(1)} u^{(1)q,p}.
\end{aligned}
\quad (4.63)
$$

$X_{rs}^{(3)}$, etc., are evaluated in an analogous way. If one considers the linear theory as a first approximation, the second approximation plainly corresponds to the assumption

$$u_r = u_r^{(1)} \theta + u_r^{(2)} \frac{\theta^2}{2} \, ,$$

$$X_r = X_r^{(1)} \theta + X_r^{(2)} \frac{\theta^2}{2} \, .$$

(4.64)

It is clear that such a theory presumes the knowledge of the terms of the power series expressing the thermodynamic potential up to and including the third order. Specifically, put

$$z_1 = I_1 \, , \qquad z_2 = I_2 \, , \qquad z_3 = D \, , \qquad (4.65)$$

$$
\left.
\begin{aligned}
J^{(1)} &= \sum_i \left(\frac{\partial J^{(D)}}{\partial z_i} \right)^{(0)} z_i^{(1)} \, , \\[2mm]
J^{(2)} &= \frac{1}{2} \left[\sum_{i,h} \left(\frac{\partial^2 J^{(D)}}{\partial z_i \, \partial z_h} \right)^{(0)} z_i^{(1)} z_h^{(1)} + \sum_i \left(\frac{\partial J^{(D)}}{\partial z_i} \right)^{(0)} z_i^{(2)} \right] , \\[2mm]
J^{(3)} &= \frac{1}{3} \left[\sum_{i,h,l} \left(\frac{\partial^3 J^{(D)}}{\partial z_i \, \partial z_h \, \partial z_l} \right)^{(0)} z_i^{(1)} z_h^{(1)} z_l^{(1)} + 3 \sum_{i,h} \left(\frac{\partial^2 J^{(D)}}{\partial z_i \, \partial z_h} \right)^{(0)} z_i^{(2)} z_h^{(1)} + \right. \\[2mm]
&\qquad \left. + \sum_i \left(\frac{\partial J^{(D)}}{\partial z_i} \right)^{(0)} z_i^{(3)} \right] ;
\end{aligned}
\right\}
$$

(4.66)

then

$$
\left.
\begin{aligned}
l^{(0)} &= \frac{\partial J^{(1)}}{\partial z_1^{(1)}} + \frac{\partial J^{(1)}}{\partial z_3^{(1)}} = 0 \, , \\[2mm]
m^{(0)} &= \frac{\partial J^{(1)}}{\partial z_1^{(1)}} - \frac{1}{2} \frac{\partial J^{(1)}}{\partial z_2^{(1)}} \, , \\[2mm]
n^{(0)} &= - 2 \frac{\partial J^{(1)}}{\partial z_2^{(1)}} \, ;
\end{aligned}
\right\}
$$

(4.67)

$$
\left.
\begin{aligned}
l^{(1)} &= \frac{\partial J^{(2)}}{\partial z_1^{(1)}} + \frac{\partial J^{(2)}}{\partial z_3^{(1)}} + \left[\frac{\partial J^{(1)}}{\partial z_2^{(1)}} + \frac{\partial J^{(1)}}{\partial z_3^{(1)}} \right] z_1^{(1)} \, , \\[2mm]
m^{(1)} &= \frac{\partial J^{(2)}}{\partial z_1^{(1)}} - \frac{1}{2} \frac{\partial J^{(2)}}{\partial z_2^{(1)}} + \frac{\partial J^{(1)}}{\partial z_2^{(1)}} z_1^{(1)} \, , \\[2mm]
n^{(1)} &= - 2 \frac{\partial J^{(2)}}{\partial z_2^{(1)}} \, ;
\end{aligned}
\right\}
$$

(4.68)

$$
\begin{aligned}
l^{(2)} &= \frac{\partial J^{(3)}}{\partial z_1^{(1)}} + \frac{\partial J^{(3)}}{\partial z_3^{(1)}} + \frac{\partial J^{(1)}}{\partial z_2^{(1)}} z_1^{(2)} + \frac{\partial J^{(1)}}{\partial z_3^{(1)}} z_3^{(2)} + \\[2mm]
&\qquad + 2 \left[\frac{\partial J^{(2)}}{\partial z_2^{(1)}} + \frac{\partial J^{(2)}}{\partial z_3^{(1)}} \right] z_1^{(1)} \, .
\end{aligned}
$$

(4.69)

§ 7. The second-order theory of Rivlin

I have presented rather explicitly the equations of a second-order theory for the determination of X_{rs} so as to make an interesting comparison with a method proposed by RIVLIN [2] in 1953. According to (3.10) let us put

$$\bar{\bar{I}}_1(\nu) = \bar{I}_1(g) - 3, \quad \bar{\bar{I}}_2(\nu) = \bar{I}_2(g) - 2\,\bar{I}_1(g) + 3,$$

$$\bar{\bar{I}}_3(\nu) = \bar{I}_3(g) - \bar{I}_2(g) + \bar{I}_1(g) - 1\,. \tag{4.70}$$

Considering isothermal displacements from an unstressed state and using for X_{rs} the expressions (3.11), RIVLIN proposes to assume as an approximation that

$$J = a_1 \bar{\bar{I}}_2(\nu) + a_2 \bar{\bar{I}}_1^2(\nu) + a_3 \bar{\bar{I}}_1(\nu)\,\bar{\bar{I}}_2(\nu) + a_4 \bar{\bar{I}}_1^3(\nu) + a_5 \bar{\bar{I}}_3(\nu)\,, \tag{4.71}$$

supposing the powers of $u_{r,s}$ of degree higher than the second to be negligible in the successive developments. $a_1, \ldots, a_5$ are constant coefficients. Putting

$$e_{ih} = u_{i,h} + u_{h,i}\,, \quad \alpha_{ih} = u_{i,s}\,u_h^{,s}\,, \quad \alpha = u_{i,s}\,u^{i,s}\,,$$

$$E_{ih} = e_{i+1\,h+1}\,e_{i+2\,h+2} - e_{i+1\,h+2}\,e_{i+2\,h+1}\,, \quad E = E_s^{,s}\,, \tag{4.72}$$

neglecting the terms of degree greater than 2, from (3.11), (4.70), (4.71), (4.72) we obtain

$$X_{rs} = X'_{rs} + X''_{rs}\,, \tag{4.73}$$

with

$$X'_{rs} = 2\left[a_1\,e_{rs} - 2(a_1 + 2a_2)\,\varDelta\,\delta_{rs}\right]\,, \tag{4.74}$$

$$X''_{rs} = 2\left\{-(4a_2 - 2a_3 + a_1)\,\varDelta\,e_{rs} + a_1\,\alpha_{rs} + (a_1 - a_5)\,E_{rs} -\right.$$
$$\left. - \delta_{rs}[(a_1 + 2a_2)\,\alpha + (a_1 + a_3)\,E + 2(6a_4 + 2a_3 - a_1 - 2a_2)\,\varDelta^2]\right\}\,.$$

Since in the linear case a_3, a_4, a_5 must be supposed equal to zero, it is evident that a_1, a_2 are expressible in terms of the Lamé coefficients:

$$a_1 = -\frac{\mu}{2}\,, \quad a_2 = \frac{\lambda + 2\mu}{8}\,. \tag{4.75}$$

Equations (4.51), written without the parameter θ and neglecting the terms of the third degree, become

$$[(1 + \varDelta)\,\delta_{hs} - u_{s,h}]\,X_r^{',h,s} + X_{rs}^{'',s} = k^* F_r\,,$$
$$[(1 + \varDelta)\,\delta_{hs} - u_{s,h}]\,X_r^{',h}\,N_*^s + X_{rs}''\,N_*^s = f_r^*\,. \tag{4.76}$$

In RIVLIN's method, first of all the solution of (4.72), (4.74), (4.75), (4.76) is determined, the non-linear terms being neglected. Let (u_i', ξ_{ih}')

be such a solution, and let $\varDelta'$, ξ''_{rs} be the expressions of $\varDelta$, X''_{rs} corresponding to u'_i. Let us denote by $k^* F'_r$, f'_r the forces to which (4.76) give rise if one identifies u_i with u'_i. Plainly,

$$
\begin{aligned}
k^* F'_r &= [(1 + \varDelta')\, \delta_{hs} - u'_{s,h}]\, \xi'^{,h,s}_r + \xi''^{,s}_{rs}\,, \\
f'_r &= [(1 + \varDelta')\, \delta_{hs} - u'_{s,h}]\, \xi'^{,h}_r N^s_* + \xi''_{rs} N^s_*\,.
\end{aligned}
\tag{4.77}
$$

Then RIVLIN considers the equations obtained from (4.76) by substituting for $k^* F_r$, f^*_r the differences $k^* (F_r - F'_r)$, $f^* - f'_r$ and then neglecting the non-linear parts of the left-hand members. Thus, equations (4.76) become

$$
\begin{aligned}
X'^{,s}_{rs} &= k^* (F_r - F'_r) = -[\varDelta'\, \xi'^{,s}_{rs} - u'_{s,h}\xi'^{,h,s}_r + \xi''^{,s}_{rs}]\,, \\
X'_{rs} N^s_* &= f^*_r - f'_r = -[\varDelta'\, \xi'_{rs} - u'_{s,h}\xi'^{,h}_r + \xi''_{rs}]\, N^s_*\,.
\end{aligned}
\tag{4.78}
$$

If u'''_i, ξ'''_{rs} is the solution of equations (4.74, 1), (4.78), then the solution of equations (4.74), (4.76) is expressed by the formulae

$$
u_r = u'_r + u''_r\,, \qquad X_{rs} = \xi'_{rs} + \xi''_{rs} + \xi'''_{rs}\,,
\tag{4.79}
$$

if one neglects terms of the third degree.

RIVLIN applied the above method to the problems of extension and torsion.

It is possible to recognize that the results of RIVLIN's method coincide with those obtained by the method explained in the last section if one neglects the third power of θ.

Therefore, possible theorems of existence and uniqueness or expansion of the solutions of equations (4.51) in power series in θ, which one may expect according to STOPPELLI's papers [see Chap. V], may give a theoretical position to RIVLIN's method (see also an observation of TRUESDELL [2]), which may be applied also to calculate u_r, X_{rs} to an order of approximation greater than the second.

Chapter V

Analytical Problems Regarding the Fundamental Equations of Isothermal Static Elasticity

§ 1. Preliminaries

In the last chapter necessary conditions for the existence of derivatives of the solution of the basic problem of static elasticity with respect to a paramater θ, to which the external forces are assumed proportional, have been exhibited. It has been seen that if u_r has derivatives of arbitrarily high order with respect to θ at $\theta = 0$, it is possible, formally, to construct the power series

$$u_r = \sum_{h=1}^{\infty} u_r^{(h)} \frac{\theta^{(h)}}{h!} , \qquad (5.1)$$

where $u_r^{(h)}$ is the solution of a linear problem analogous to that of the linear theory of elasticity, with certain known terms expressed as functions of $u_r^{(1)}, u_r^{(2)}, \ldots, u_r^{(h-1)}$ for $h \geqslant 2$, while $u_r^{(1)}$ is simply the solution of the basic problem for the case of small strain. Therefore, if the expansion (5.1) is justified, the integration of the fundamental problem of large strain may be reduced to one of determining the solutions of certain linear problems of small deformation problems which in concrete cases are finite in number, since for practical applications some finite number of terms in (5.1) will be taken. Thus it is interesting to establish theorems of existence and uniqueness of the solution of the basic analytical problem and to see if and when it is possible to express them by power series in θ. In this field the papers of STOPPELLI [2, 3, 4, 5, 6, 7] are fundamental and are the basis of what is presented in the first five sections of this chapter. STOPPELLI's results solve the question completely in the case when $A \neq 0$ [see (4.31)], the orientation of C^* then being principal of infinite order, while if $A = 0$ they allow us to see if an expansion of type (5.1) is permissible in certain interesting cases.

§ 2. Fundamental hypotheses. Statement of the theorems of existence and uniqueness in the case of a principal orientation of infinite order

1. The first five sections of this chapter refer to the case of isothermal equilibrium subject to known external forces [absence of constraints]. The fundamental hypotheses are the following:

α) C^* is a configuration of natural equilibrium; that is, C^* is in equilibrium with $Y_{rs} \equiv 0$ and without external forces;

β) the boundary Σ^* of C^* has Hölder-continuous principal curvatures with exponent σ, and it is possible to cover it by a finite number of regions $\Sigma_*^{(p)}$, each of which is parametrically representable on a circle $C^{(p)}$;

γ) the vectors $k^* F, f^*$ of the external forces are in equilibrium on C^*; i.e., they satisfy the equalities

$$\int_{C^*} k^* F \, dC^* + \int_{\Sigma^*} f^* \, d\Sigma^* = 0 , \qquad (5.2)$$

$$\int_{C^*} OP^* \times k^* F \, dC^* + \int_{\Sigma^*} OP^* \times f^* \, d\Sigma^* = 0 ; \qquad (5.3)$$

δ) $k^* F_r$ are Hölder-continuous with exponent σ in C^*;

ε) f_r^* and their derivatives with respect to the parameters of a parametric representation of Σ^* are Hölder-continuous with exponent σ on Σ^*.

φ) if $u_{i,j}$ varies within a sphere $\bar{E}$ having its center at the origin and radius R, the isothermal potential $W\,(\varepsilon_{pq})$ according to (1.11) is continuous together with its derivatives up to and including the fourth order with respect to $u_{i,j}$.

2. If a function $\varphi\,(P^*)$ is said to be Hölder-continuous with exponent σ, it is meant that, if

$$q = \frac{\varphi\,(P_*) - \varphi\,(P'_*)}{|P_* \, P'_*|^\sigma} \, , \tag{5.4}$$

the quantity q is bounded when P^*, P'_* vary over C^*. The least upper bound of this quotient is the Hölder coefficient of $\varphi\,(P_*)$. Let S be the linear space of all vectors whose components satisfy the conditions

$$u_r = 0\,, \quad u_{r,r+1} - u_{r+1,r} = 0 \quad \text{(for } y_1 = y_2 = y_3 = 0) \tag{5.5}$$

and have in C^* Hölder-continuous second derivatives with exponent σ, and for the norm $||x||$ in this space take

$$||x|| = \sum_i \{ M_{ax}|u_i| + M_{ax}\,|u_{i,j}| + M_{ax}\,|u_{i,jk}| + B_{ijk} \}\,, \tag{5.6}$$

where B_{ijk} are the Hölder coefficients for the second derivatives. The following theorem of existence and uniqueness holds:

Theorem A): *If the set of external forces applied on C^* admits no axis of equilibrium $[A \neq 0]$, and if hypotheses $\alpha)$, $\beta)$, $\gamma)$, $\delta)$, $\varepsilon)$, $\varphi)$ are satisfied, then positive numbers ϱ, $\bar{\theta}$ exist such that for any value of θ for which $|\theta| < \bar{\theta}$, the basic equations*

$$\begin{aligned}
(x_{r,m}\,Y^{l,m})_{,l} &= \theta k^* F_r & \text{(in } C^*)\,, \\
x_{r,m}\,Y^{lm}\,N^*_l &= \theta f^*_r & \text{(on } \Sigma^*)\,, \\
Y_i &= -\frac{\partial W}{\partial \varepsilon_j}\,,
\end{aligned} \tag{5.7}$$

have one and only one solution which satisfies (5.5) and the conditions

$$||x|| < \varrho\,, \qquad \lim_{\theta \to 0} ||x|| = 0\,. \tag{5.8}$$

§ 3. Demonstration of Theorem A

1. To demonstrate Theorem A), following the method of STOPPELLI, it is appropriate to consider the set of equations

$$\begin{aligned}
(x_{r,m}\,Y^{lm})_{,l} &= \theta k^* \gamma_{rs} F^s & \text{(in } C^*)\,, \\
x_{r,m}\,Y^{lm}\,N^*_l &= \theta \gamma_{rs} f^{*s} & \text{(on } \Sigma^*)\,, \\
y_j &= -\frac{\partial W}{\partial \varepsilon_j}\,,
\end{aligned} \tag{5.9}$$

where γ_{rs} are constant parameters which characterize a rigid rotation and which are to be determined, if possible, in such manner that a theorem of existence and uniqueness is valid for equations (5.9).

Theorem A′): *Under the same hypotheses as for Theorem A), for any sufficiently small value of $|\theta|$ one and only one set of values of the parameters γ_{rs} exists for which equations (5.9) have one and only one solution satisfying (5.5).* Then, it is evident that if $\bar{x}_r$ is that solution, and if $\bar{\gamma}_{rs}$ are the values of γ_{rs} for which that solution exists and is unique, the set of equations (5.7) always admits the solution

$$x_r = \bar{\gamma}_{sr}\, \bar{x}^s \tag{5.10}$$

and none other.

2. To demonstrate *Theorem A′)* let us start by observing that

$$|Y_r\,(u_{i,j}) - Y_r\,(v_{i,j})| < M_0 \sum_{i,j} |u_{i,j} - v_{i,j}|, \tag{5.11}$$

where M_0 is the largest of the maxima in E [see hypothesis φ)] of $\left|\dfrac{\partial}{\partial u_{i,j}}\dfrac{\partial W}{\partial \varepsilon_p}\right|, \left|\dfrac{\partial^2}{\partial u_{i,j}\,\partial u_{q,l}}\dfrac{\partial W}{\partial \varepsilon_p}\right|$. C^* being an unstressed state of equilibrium, from (5.11) it follows that

$$|Y_r\,(u_{p,q})| < M_0 \sum_{i,j} |u_{i,j}|. \tag{5.12}$$

Then, from (5.6), (5.12),

$$\left|\int_{C^*}(u_{r,s}\,Y^{\cdot s}_{r+1} - u_{r+1,s}\,Y^{\cdot s}_r)\,dC^*\right| < C^*M_0\,||x||^2. \tag{5.13}$$

It is evident that the equations

$$\begin{aligned} (x_{r,i}\,Y^{is})_{,s} - \theta k^*\gamma_{rs}F^s &= G_r & \text{(in } C^*\text{)},\\ x_{r,i}\,Y^{is}\,N^*_s - \theta\gamma_{rs}\,f^s_* &= g_r & \text{(on } \Sigma^*\text{)},\\ Y_j &= -\dfrac{\partial W}{\partial \varepsilon_j}, \end{aligned} \right\} \tag{5.14}$$

in the domain $||x|| < R$ determine a correspondence under which to any point x of S there corresponds a sextuple G_r, g_r verifying equalities such as (5.2).

3. Put

$$A_{rs} = \int_{C^*} k^*y_r F_s\,dC^* + \int_{\Sigma^*} y_r f^*_s\,d\Sigma^*. \tag{5.15}$$

It is certainly possible to orient the reference frame T in such a manner that $A_{rs} = 0$ when $r \neq s$. I suppose that T satisfies this condition. Then the condition that the parameters γ_{rs} of the first members of equations

(5.14) give rise to sextuples satisfying relations like (5.3) is expressed by
the equations

$$\theta \left(A_{r+1\,r+1}\, \gamma_{r+2\,r+1} - A_{r+2\,r+2}\, \gamma_{r+1\,r+2} \right) = L_r \,, \tag{5.16}$$

$$\gamma_{rt}\,\gamma_s^{\cdot t} = \delta_{rs}\,, \qquad \| \gamma_{rs} \| = 1\,, \tag{5.17}$$

where

$$L_r = \int_{C*} \left(u_{r+2,\,s}\, Y_{r+1}^{\cdot s} - u_{r+1,\,s}\, Y_{r+2}^{\cdot s} \right) dC* \,. \tag{5.18}$$

It is possible to demonstrate (STOPPELLI [1]) that two positive num-
bers ϱ_1, η_1 always exist such that for any $\eta > 0$ and $< \eta_1$ one may determine another positive number $\varrho_0 \leqslant \varrho_1$ for which equations (5.16), (5.17)
admit one and only one solution satisfying the relation

$$| \gamma_{rs} - \delta_{rs} | < \eta \tag{5.19}$$

when

$$\sqrt{\sum_r L_r^2} \leqslant \varrho_0\, | \theta |\,. \tag{5.20}$$

For $L_r = 0$ this solution becomes $\gamma_{rs} = \delta_{rs}$.
Since from (5.13), (5.18) it follows that

$$\sqrt{\sum_r L_r^2} \leqslant \sqrt{3}\, C* M_0\, \| x \|^2\,, \tag{5.21}$$

one deduces that (5.20) is verified if

$$\| x \| \leqslant \sqrt{\frac{\varrho_0}{\sqrt{3}\, C* M_0}}\, | \theta |^{\frac{1}{2}}\,, \tag{5.22}$$

a relation which, since $\| x \| < R$, is certainly satisfied when

$$| \theta | \leqslant \frac{\sqrt{3}\, R^2 C* M_0}{\varrho_0}\,. \tag{5.23}$$

Therefore, if θ satisfies (5.23), one and only one sextuple G_r, g_r exists
which verifies equations such as (5.2), (5.3), for any x in a neighborhood
$\bar{J}_0$ of the point $\Omega \equiv (0, 0, 0)$ defined by (5.22). When x varies in $\bar{J}_0$, such
a sextuple determines a point x' of a space S', namely, that one of the
sextuples Ψ_1, Ψ_2, Ψ_3, φ_1, φ_2, φ_3 which verifies conditions γ), δ), ε). The
space S' is linear, complete and normal, and its norm will defined by

$$\| x' \| = \sum_i \{ M_{ax} | \psi_i | + M_{ax} | \varphi_i | + C_i + D_i \}\,, \tag{5.24}$$

where C_i is the Hölder coefficient for Ψ_i, while D_i is the largest of the
Hölder coefficients for the circles $C^{(1)}, \ldots, C^{(n)}$ of the first derivatives of
φ_i with respect to the parameters of a parametric representation of $\Sigma*$
[see hypothesis β)]. When x varies in $\bar{J}_0$, x' varies in a neighborhood $\bar{J}_0'$,

4*

of point Ω' which corresponds to Ω, and so the functional transformation

$$x' = T(x) \tag{5.25}$$

is defined. It is evident that $L_r = 0$ corresponds to the point $\Omega \equiv (0, 0, 0)$, and consequently $\gamma_{rs} = \delta_{rs}$ [see (5.16), (5.17)]. Then it follows that the corresponding Ω' of Ω according to (5.14) is defined by the sextuple

$$G_r(\Omega) = -\theta k^* F_r, \qquad g_r(\Omega) = -\theta f_r^*. \tag{5.26}$$

4. Let

$$D(x_0, x) = \lim_{t \to 0} \frac{T(x_0 + tx) - T(x_0)}{t} = \left[\frac{d}{dt} T(x_0 + tx)\right]_{t=0}, \tag{5.27}$$

$$R(x_0, x) = T(x_0 + x) - T(x_0) - D(x_0, x). \tag{5.28}$$

$D(x_0, x)$ is called the differential of $T(x)$ at the point x_0. One may demonstrate that $T(x)$ is *continuously differentiable* in $\bar{J}_0$. That means that however small a positive number τ is chosen, one may always find another one, ϱ_τ, for which the two following conditions are satisfied:

I) For any pair of points x_0, $\bar{x}_0$ of $\bar{J}_0$ satisfying the inequality

$$\| x_0 - \bar{x}_0 \| \leqslant \varrho_\tau, \tag{5.29}$$

and for any x of $\bar{J}_0$, we have

$$\| D(x_0, x) - D(\bar{x}_0, x) \| \leqslant \tau \| x \| ; \tag{5.30}$$

II) for any x_0 of $\bar{J}_0$, if $\|x\| \leqslant \varrho_\tau$, we have

$$\| R(x_0, x) \| \leqslant \tau \| x \| . \tag{5.31}$$

For brevity I omit the demonstration, which may be found in STOP-PELLI's paper [2]. I recall only that it consists in showing the possibility of finding a positive number h, independent of θ, for which

$$\| D(x_0, x) - D(\bar{x}_0, x) \| < h|\theta|^{-\frac{1}{2}} \| x_0 - \bar{x}_0 \| \cdot \| x \| ,$$
$$\| R(x_0, x) \| \leqslant h|\theta|^{-\frac{1}{2}} \| x_0 - \bar{x}_0 \| \cdot \| x \| . \tag{5.32}$$

Then, if one assumes that

$$\varrho_\tau = \frac{\tau}{h} |\theta|^{\frac{1}{2}}, \tag{5.33}$$

(5.30) and (5.31) hold.

5. It is easily seen that the linear transformation

$$D(\Omega, \delta x) = \delta x', \tag{5.34}$$

which is deduced by differentiation of (5.25), coincides with the fundamental linear system of static elasticity, namely,

$$
\begin{aligned}
Y_{rs}^{(1),s} &= \delta\,G_r && \text{(in } C^*) , \\
Y_{rs}^{(1)} N_*^s &= \delta\,g_r && \text{(on } \Sigma^*) , \\
Y_{rs}^{(1)} &= \left[\frac{d}{dt}\,Y_{rs}(\Omega + tx)\right]_{t=0} .
\end{aligned}
\right\}
\tag{5.35}
$$

Let us assume that the set of equations (5.35) has one and only one solution $u_r^{(1)}$ satisfying (5.5) [see Section 6]. Since $u_r^{(1)}$ are solutions of equations of elliptic type, they have Hölder-continuous second derivatives with exponent σ. By known theorems of functional analysis [MIRANDA], that fact is sufficient to assure that (5.25) is locally invertible from a neighborhood of Ω to one of

$$\Omega' \equiv T(\Omega) . \tag{5.36}$$

Specifically, if $|\theta|$ is sufficiently small, it is possible to determine a positive number r such that for any x' satisfying the inequality

$$\| x' - \Omega' \| \leqslant r\,|\theta|^{\frac{1}{2}} , \tag{5.37}$$

there exists one and only one solution of equation (5.25) defined in neighborhood J_0 of Ω. By reason of (5.22) for $x' = 0$, (5.37) becomes

$$|\theta|\sum_i \{M_{ax}\,|F_i| + M_{ax}\,|f_i^*| + C_i + D_i\} \leqslant r\,|\theta|^{\frac{1}{2}} , \tag{5.38}$$

which is satisfied if $|\theta|$ is sufficiently small. That means that the point $x' = 0$ satisfies (5.36) when $|\theta|$ is sufficiently small, and therefore the equation

$$T(x) = 0 \tag{5.39}$$

admits one and only one solution. Thus Theorem A') is demonstrated. Theorem A) then follows.

§ 4. On the expansion of the solution of the basic set of equations in a power series in θ when there is no axis of equilibrium

1. All necessary elements have been assembled for demonstration of the following **Theorem B)**: *The solution of the fundamental set of equations* (5.7) *is a holomorphic function of* θ *if* $|\theta|$ *is sufficiently small.*

Thence one deduces the possibility of expressing u_r by a power series in θ, foreseen by SIGNORINI [4]. To demonstrate Theorem B), together with hypothesis φ), let us consider a further one, φ'): the isothermal elastic potential $W(\varepsilon_{pq})$ is a holomorphic function of the ε_{pq}

when they vary in a complex space subject to the condition

$$|\varepsilon_j| \leqslant \bar{R} . \tag{5.40}$$

Then we may write

$$|u_{i,j}| \leqslant R' . \tag{5.41}$$

Certainly

$$|\varepsilon_{ij}| \leqslant 2R' + 3R'^2 . \tag{5.42}$$

Therefore, if one assumes that

$$R' \leqslant \frac{1}{3}[\sqrt{1+3\bar{R}} - 1], \tag{5.43}$$

it results that $|\varepsilon_j| \leqslant \bar{R}$ and, then, by hypothesis φ'), $W(\varepsilon_j)$ and all its derivatives are holomorphic functions of $u_{i,j}$ if these satisfy (5.41). Further, it is appropriate to observe that the Jacobian determinant of the left-hand members of equations (5.16), (5.17) is different from zero (STOPPELLI [3]); therefore the solutions of those equations, namely the γ_{rs}, are holomorphic functions of θ in a circle Γ if the L_r are.

2. Let us consider the inverse transformation of (5.34):

$$\delta x = \Delta'(\delta x') . \tag{5.44}$$

Since Δ' is a linear operator, a positive constant h exists such that

$$\frac{||\delta x||}{||\delta x'||} \leqslant \bar{h} . \tag{5.45}$$

Let us suppose that $\delta G_r, \delta g_r$ occurring in equations (5.35) are holomorphic functions of θ when θ varies in Γ and that the norm of the element defined by them in S_0 is bounded in Γ. It follows that the expansions

$$\delta G_r = \sum_{n=0}^{\infty} \frac{\delta G_r^{(n)}}{n!} \theta^n , \qquad \delta g_r = \sum_{n=0}^{\infty} \frac{\delta g_r^{(n)}}{n!} \theta^n \tag{5.46}$$

are legitimate in a certain circle Γ_1.

Then from (5.46) it follows that the equations of the classical linear theory of elasticity corresponding to the element $\delta G_r^{(n)}, \delta g_r^{(n)}$, namely,

$$\left.\begin{aligned}
Y_{rs}^{(1),s} &= \delta G_r^{(n)} &&\text{(in } C^*) , \\
Y_{rs}^{(1)} N_*^s &= \delta g_r^{(n)} &&\text{(on } \Sigma^*) , \\
Y_{rs}^{(1)} &= \left[\frac{d}{dt} Y_{rs}(\Omega + tx)\right]_{t=0} ,
\end{aligned}\right\} \tag{5.47}$$

admit one and only one solution $u_r^{(n)}$ satisfying (5.5), having Hölder-continuous second derivatives with exponent σ, and satisfying the inequality

$$|u_i^{(n)}| \leqslant \bar{h} \sum_i \{M_{ax} |\delta G_i^{(n)}| + M_{ax} |\delta g_i^{(n)}| + C_i^{(n)} + D_i^{(n)}\} . \tag{5.48}$$

From (5.48) one easily deduces that the series $\sum\limits_{n=0}^{\infty} \frac{u_r^{(n)}}{n!} \theta^n$ is uniformly convergent in Γ, and therefore the solution of (5.35) is a holomorphic function of θ in Γ.

3. Let

$$\mathfrak{F}(x) = x - \varDelta \, [T(x)] , \tag{5.49}$$

and let J_0' be the neighborhood of $\varOmega$ defined by the inequality

$$\|x\| < \varrho \, |\theta|^{\frac{1}{2}} . \tag{5.50}$$

From (5.32) follows the possibility of determining ϱ in such manner that

$$\|\mathfrak{F}(z) - \mathfrak{F}(z')\| < K \, \|z - z'\| , \tag{5.51}$$

where $K < 1$.

It is evident that (5.39) is equivalent to the equation (5.52), namely

$$z = \mathfrak{F}(z) . \tag{5.52}$$

If it is assumed that

$$z_0 = \varOmega , \quad z_1 = \mathfrak{F}(z_0) , \quad z_2 = \mathfrak{F}(z_1),\dots, z_{n+1} = \mathfrak{F}(z_n),\dots, \tag{5.53}$$

one may demonstrate (STOPPELLI [3]) the following properties:
a) the elements $z_0, z_1,\dots$ belong to J_0';
b) the sequence $\{z_n\}$ converges to an element of S.

Further, the element to which $\{z_n\}$ converges satisfies (5.52) and is unique, as follows from (5.53) for $n \to \infty$.

4. In fact, the typical member of (5.53) is

$$D\,(\varOmega, z_{n+1}) = D\,(\varOmega, z_n) - T\,(z_0) , \tag{5.54}$$

equivalent to

$$D\,(\varOmega, z_{n+1}) = - \sum_{j=1}^{n} T\,(z_j) . \tag{5.55}$$

This means that with

$$U_r\,(x) = (x_{r,i}\,Y^{is})_{,s} , $$
$$V_r\,(x) = x_{r,i}\,Y^{is}\,N_s^* , \tag{5.56}$$

z_{n+1} is the solution of equations (5.35) when the quantities $\delta A_r^{(n)}$, $\delta a_r^{(n)}$, given by the equalities

$$\delta A_r^{(n)} = - \sum_{j=0}^{n} U_r\,(z_j) + \theta k^* \sum_s F_s \sum_{j=0}^{n} \gamma_{rs}\,(z_j) , $$
$$\delta a_{r i}^{(n)} = - \sum_{j=0}^{n} V_r\,(z_j) + \theta \sum_s f_s^* \sum_{j=0}^{n} \gamma_{rs}\,(z_j) \tag{5.57}$$

are substituted into (5.35) in place of δG_r, δg_r, respectively. In particular,

$$\delta A_r^{(0)} = \theta k * F_r , \qquad \delta a_r^{(0)} = \theta f_r^* , \tag{5.58}$$

from which it results that $\delta A_r^{(0)}$, $\delta a_r^{(0)}$ are holomorphic functions of θ. Consequently, by what was said for the solution of (5.53), z_1 is also a holomorphic function of θ, and so also is $L_r^{(1)}$, by reason of the hypothesis φ') made for $W (\varepsilon_{pq})$; therefore, finally, so is $\gamma_{rs} (z_1)$. In conclusion, the sequence of elements $z_1, z_2, \ldots z_n$ is a sequence of holomorphic functions and converges toward a holomorphic function if $|\theta|$ is sufficiently small. Then, γ_{rs} being holomorphic, from (5.10) it follows that also the u_r are holomorphic functions of θ.

§ 5. On the existence of solutions and the possibility of expanding them in power series in θ when there is an axis of equilibrium

1. In previous sections it has been seen that if the set of external forces applied to C^* has no axis of equilibrium, the fundamental equations admit one and only one solution satisfying (5.5), and that solution may be expressed by a power series in θ if $|\theta|$ is sufficiently small.

But if the external forces admit an axis of equilibrium, the successive systems of fundamental equations may admit no solution [Chap. IV], and power series expansion of the solution of the basic problem may be impossible. In this case one is doubtful about the meaning of the classical linear theory if one wishes to interpret it as the first term of a power series of θ. One may even be doubtful about the very existence of solutions of the fundamental equations of static elasticity, apart from the question of power series expansion, when there is an axis of equilibrium.

This question, too, has been studied by F. Stoppelli. I will limit myself to stating the most interesting theorems established by him, refering to the original papers for the proofs. By these theorems it results that in some cases the solution of the basic problem exists, is unique and may be expanded in power series in θ when $|\theta|$ is sufficiently small.

A vector $S (P^*, \theta)$ will be said to belong to class C if it is zero at 0 and has Hölder-continuous second derivatives for any value of θ of sufficiently small modulus, while its first derivatives with respect to y_r are continuous functions of θ. Further, the vector $S (P^*, \theta)$ will be said to *vanish in the strong sense* for $\theta = 0$ if it is infinitesimal together with the first and second derivatives with respect to y_r and the Hölder coefficients of the second derivatives when $\theta = 0$.

I shall say that the vector $S (P^*, \theta)$ is of class C' if it vanishes in the strong sense for $\theta = 0$ and belongs to class C. The following theorems hold (Stoppelli [5]):

Theorem C): *A necessary condition that the fundamental set* (5.7) *admit a solution of class C' is that C^* be a principal orientation of the first order.*

Theorem D): *If the orientations of C^* for which the set of external forces applied to C^* is in equilibrium are all of the first order, a necessary condition for the existence of a solution of class C' of the fundamental set of equations is that C^* be a principal orientation of the second order.*

From Theorems C), D) it follows that a necessary condition for the existence of a solution of class C' of the basic equations of static elasticity is that the solution of the linear equations of the classical theory satisfies the integrability condition (4.30) for $s = 2$.

2. Let the vector $w_s \equiv (w_{1s}, w_{2s}, w_{3s})$ be the solution of the equations

$$
\left.
\begin{aligned}
Y_{rl}^{(1),l} &= k^* F_r \left[(1 - \delta_{r3}) \, \delta_{s1} + \delta_{r3} \delta_{s3} \right] + \\
&\qquad + (-1)^{r+1} (1 - \delta_{r3}) \, k^* F_{3-r} \delta_{s2} \quad (\text{in } C^*), \\
Y_{rl}^{(1)} \, N_*^l &= f_r^* \left[(1 - \delta_{r3}) \, \delta_{s1} + \delta_{r3} \delta_{s3} \right] + \\
&\qquad + (-1)^{r+1} (1 - \delta_{r3}) \, f_{3-r}^* \delta_{s2} \quad (\text{on } \Sigma^*), \\
Y_{rl}^{(1)} &= \left[\frac{d}{dt} Y_{rs} (\Omega + tx) \right]_{t=0},
\end{aligned}
\right\}
\tag{5.59}
$$

which satisfies the conditions

$$
w_{rs}(0) = 0, \qquad w_{r+2s,\,r+1} - w_{r+1s,\,r+2} = 0. \tag{5.60}
$$

Let

$$
A_r(g) = \int_{C^*} g k^* F_r dC^* + \int_{\Sigma^*} g f_r^* d\Sigma^* \tag{5.61}
$$

and

$$
\left.
\begin{aligned}
d_1 &= A_1(w_{12}) + A_2(w_{22}) = A_2(w_{11}) - A_1(w_{21}), \\
d_2 &= - A_1(w_{11}) - A_2(w_{21}) + A_2(w_{12}) - A_1(w_{22}), \\
d_3 &= - A_1(w_{23}) + A_2(w_{13}), \quad d_4 = - A_2(w_{23}) - A_1(w_{13}).
\end{aligned}
\right\}
\tag{5.62}
$$

There is a theorem by which it is possible to decide if and when an orientation of C^* is principal of the first order. This theorem, helpful in determining the existence of solutions of the fundamental equations when there is an axis of equilibrium, the necessary conditions of the theorems C), D) being satisfied, is the following (STOPPELLI [6]):

Theorem E): *A necessary and sufficient condition for an orientation of C^* to be principal of the first order is that one of the three relations*

a) $d_2 + d_4 \neq 0,$

b) $d_2 + d_4 = 0,$ $d_1 \neq 0,$

c) $d_1 = 0,$ $d_2 = - d_4 \neq 0$

be verified.

3. The following theorems hold when C^* has only one axis of equilibrium and when its principal orientations of the first order are finite in number and when C^* is one of them (STOPPELLI [6]).

Theorem F): *If condition* a) *of Theorem E) is satisfied, a solution of the set of the basic equations belonging to class C' exists if $|\theta|$ is sufficiently small. It is unique and expressible by a power series in θ.*

Theorem G): *If conditions* b) *of Theorem E) are verified and if $|\theta|$ is sufficiently small, one of the following cases necessarily holds for the set of fundamental equations:* 1°) *they admit one or two solutions of class C' expressible by power series in θ;* 2°) *they admit two solutions, but not belonging to class C, only if $\theta > 0$ [$\theta \leqslant 0$]. These solutions are expressible by power series in $\theta^{\frac{1}{2}}$ [$(-\theta)^{\frac{1}{2}}$];* 3°) *they admit no solution vanishing in the strong sense for $\theta = 0$.*

Theorem H): *If conditions* c) *of Theorem E) are satisfied and if $|\theta|$ is sufficiently small, one of the following cases necessarily holds for the basic equations:* 1°) *they admit one, two or three solutions of class C' which are expressible by power series in θ;* 2°) *they admit only one solution of class C', but it may be expanded in a power series in $\theta^{\frac{1}{3}}$;* 3°) *they admit three solutions vanishing in the strong sense for $\theta = 0$, but only one of them belongs to class C and is expressible by a power series in θ while the other two are valid only for $\theta > 0$, [$\theta \leqslant 0$] and may be expanded in power series in $\theta^{\frac{1}{2}}$ [$(-\theta)^{\frac{1}{2}}$].*

In connection with Theorems G), H) it is interesting to observe that the solution of the fundamental equations (5.7) may appear or disappear when the sign of all external forces changes.

4. The case when the external forces are all parallel is not included in the previous cases. Under this hypothesis only one axis of equilibrium exists, or the set of external forces is astatic (i.e. $a_{rs} = 0$, (4.29)]. The following theorems hold (STOPPELLI [7]):

Theorem I): *If the external forces are parallel, and if $|\theta|$ is sufficiently small, equations (5.7) admit one and only one solution of class C' satisfying the condition*

$$(u_{2,1} - u_{1,2})_0 = 2c(\theta) \tag{5.63}$$

where $c(\theta)$ is an arbitrary continuous function of θ which is infinitesimal with θ. The functions $u_r(y, \theta)$ are expressible by power series in θ if $c(\theta)$ is.

Theorem L): *If the system of parallel external forces is astatic, and if the straight line for O^*, parallel to the forces, is an isolated principal axis of the*

first order, then equations (5.7) *admit one and only one solution of class C'
satisfying* (5.63), *when* $|\theta|$ *is sufficiently small. The functions* $u_r(y, \theta)$ *are
expressible by power series in* θ *if* $c(\theta)$ *is.*

§ 6. Theorems of existence and uniqueness for the linear equations of isothermal static elasticity

1. From the previous sections it is clear that theorems of existence and uniqueness for the fundamental equations of static elasticity are subordinated to analogous theorems for the corresponding linear system of classical elasticity. On the other hand this problem in linear elasticity is interesting for its own sake, independently of its relation to the non-linear problem, since the problem of elastic equilibrium is often reduced to the linear one, to study a great number of practical problems. This question is the subject of many papers, especially concerning the case when the displacements are assigned on the boundary. For this case, demonstrations of the existence theorem when the boundary is assumed free of singularities are found, for example, in papers of LAURICELLA and LICHTENSTEIN. A theorem of existence has been given by KORN in the case when the external forces are known everywhere on the boundary. More recent is the attainment of an existence theorem for the difficult mixed boundary value problem, i.e., for the case in which the displacements are assigned on a portion of the boundary and the forces are known on the remainder.

Such a theorem has been established, for the first time, by G. FICHERA [4] and is contained in a paper giving theorems of existence and uniqueness for the three fundamental linear boundary value problems, under conditions of greater generality than those considered by previous authors.

In the non-linear case existence theorems have been established by STOPPELLI only subject to the hypothesis that the external forces are known everywhere and subordinately to the existence of the solution of the corresponding linear problem. It is desirable that analogous theorems of existence be established when constraints are present. Perhaps that is possible by following STOPPELLI's methods, but surely the theorems of existence of the corresponding linear cases are necessary.

The procedures used by FICHERA to demonstrate existence theorems are inspired by general methods of integration due to M. PICONE [2, 4] and based on a functional interpretation of BETTI's reciprocity theorem.

The results of FICHERA's studies concern a set of differential equations where the unknown functions are the components of displacement, similar to those for the linear static problem of a homogeneous, isotropic elastic body, but allowing more general conditions of variability for the coefficients.

2. Let us suppose that boundary Σ^* of C^* consists in a finite number of portions of regular surfaces, and let $\bar{C}^*$ be the unbounded domain whose boundary is Σ^*. A vector is said to be *biregular* in C^* if it is continuous in C^* together with its first derivatives. Let U be the set of all biregular vectors in C^*, and let $\bar{U}$ be the set of vectors biregular in $\bar{C}^*$ and infinitesimal at infinity. The three fundamental analytic problems related to the linear problem of elastic equilibrium correspond to three kinds of boundary conditions[1]:

Problem 1) $\qquad u = \bar{u}\,(P^*) \qquad$ (on Σ^*) ;

Problem 2) $\quad L\,(u) = f^*(P^*) \qquad$ (on Σ^*) ;

$$
\left.
\begin{array}{ll}
u = \bar{u}\,(P^*) & (\text{on } \Sigma_1^*) \\[2mm]
L\,(u) = f^*\,(P^*) & (\text{on } \Sigma_2^*)
\end{array}
\right\} \quad (\Sigma^* = \Sigma_1^* + \Sigma_2^*) ,
$$

Problem 3)

where $\bar{u}\,(P^*)$ and $f^*(P^*)$ are assigned and $L\,(u)$ is a given differential operator. Constraints are present in problems 1) and 3) but, as is evident, they are constraints yielding assigned displacements. The constraint of clamping is included, but a unilateral[2] supporting constraint plainly is not. The following theorem of uniqueness holds (FICHERA [4]): *Subject to the well known inequalities satisfied by the Lamé constants, the solution of problems 1), 2) and 3) is unique in the class $\bar{U}$.*

Indeed, it is unique in the class U in Problem 1); is determined to within a rigid displacement of the body in Problem 2); is unique if three non-collinear points belong to Σ_1^* in Problem 3).

3. G. FICHERA demonstrates the existence theorems under the hypothesis that body forces are absent, since it is evident that the general problem usually may be reduced to this case [for example, when the body forces are Hölder-continuous in C^*]. Therefore, his theorems concern the differential equation

$$
\Delta_2\, u + \frac{\lambda + \mu}{\mu}\,\text{grad div } u = 0 \qquad (\text{in } C^*) ,
$$

$$
\frac{\lambda + \mu}{\mu} > \frac{1}{3} ,
$$

(5.64)

---

[1] Hereafter, in this chapter and in the next three [except Chap. VII, § 7] only problems of the linear theory will be considered. They correspond to the first term of the expansion (5.1), but for simplicity the index (1) is omitted from symbols for functions.

[2] In calling a constraint *unilateral* I mean that inequalities are necessary for its analytical expression. For example, a constraint of support may be unilateral.

to which the boundary conditions of problems 1), 2) and 3) are to be adjoined. The proofs are based on some inversion theorems which yield an integral expression of the solution by use of Somigliana's matrix. In summary, these theorems are:

Problem 1): *A sufficient condition for existence is that the vector $\bar{u}$ assigned on Σ^* be the trace there of a vector $w(P^*)$ defined in C^*, having square-integrable divergence and curl and being uniformly integrable on the boundaries of a sequence of regular domains approaching C^*.*

The above condition for $\bar{u}$ is very general. In particular, it is satisfied if $\bar{u}$ is the trace on Σ^* of a vector whose first derivatives have integrable norm on Σ^*, or is the trace of a biregular vector in C^*.

Problem 2): Let V be the variety of vectors $f^*(P^*)$ defined on Σ^* and such that corresponding to each of them, there exist three continuous vectors $f^{(i)}(P^*)$ defined in C^* which have continuous first derivatives in $C^* - \Sigma^*$ and which satisfy the conditions

$$f_r^*(P^*) = f^{(r)}(P^*) \cdot N^* \qquad \text{(on } \Sigma^*\text{)} , \qquad \text{a)}$$

$$\int_{\Sigma'} f^{(i)} \cdot N^* d\Sigma' = 0$$
$$\qquad\qquad\qquad (i,j=1,2,3) , \qquad \Bigg\} \quad \text{b)}$$
$$\int_{\Sigma'} [y_j f^{(i)} \cdot N^* - y_i f^{(j)} \cdot N^*] \, d\Sigma' = 0$$

where condition b) holds on the boundary Σ' of any regular domain C' contained within C^*.

Then a necessary and sufficient condition that problem 2) has a solution in a certain class C_u is that the assigned vector $f^*(P^*)$ on Σ^* belongs to the variety V.

The class C_u is that of the vectors which are solutions of equation (5.64) and which, together with the vectors of Somigliana's matrix, satisfy an integral equality like Betti's reciprocity theorem. In particular, the class C_u contains the biregular vectors in C^* which satisfy (5.64).

Problem 3): *A sufficient condition that Problem 3) admits a solution belonging to a certain class C'_u is that the vector $f^*(P^*)$, assigned on Σ_2^*, be the trace on Σ_2^* of a vector which is defined on the whole of Σ^* and belongs to the variety V.*

The class C'_u is that of the vectors which satisfy (5.64), are zero on Σ_1^*, are continuous in $C^* - \Sigma_2^*$ with continuous first and second derivatives in $C^* - \Sigma_2^*$ and are such that together with some particular solutions of the linear, homogeneous, fundamental problem constructed by vectors of Somigliana's matrix they satisfy certain integral relations of reciprocity. In particular, the biregular vectors in C^* satisfying (5.64) and vanishing on Σ_1^* belong to class C'_u. Thus the existence theorem for problem 3) is established under the hypothesis that the body is clamped on $\Sigma_1^* [\bar{u}(P^*) = 0$

on Σ_1^*]. It is interesting to observe that in passing on Σ^* from Σ_1^* to Σ_2^* across the line l^* which divides the clamped portion from the one where the surface forces are assigned, the stress need not be continuous. Certainly, *continuity fails* if the assigned surface forces on Σ_2^* do not satisfy a certain integral condition, corresponding to the form of Σ_2^*. Therefore, it may be presumed that if the stress corresponding to a certain set of surface forces on Σ_2^* is continuous across l^*, it *may* become discontinuous by changing the surface forces even only on a portion of Σ_2^* which is distant from l^*. An analytic integral condition established by G. FICHERA [5] is to be understood in this sense.

Chapter VI

Inequalities for the Equilibrium of Slightly Deformable Elastic Bodies

§ 1. Preliminaries

In general, the boundary-value problem of static elasticity is difficult to solve even in the linear theory. I shall take up the matter in the next chapter, but now I call attention to the fact that it certainly may be useful, and often it is sufficient, to establish certain inequalities for the components of stress and displacement. After all, a method of integration generally gives the values of the unknown functions only to within some error. For engineering use it is often sufficient to make sure that the magnitude of a given component of stress is less than a certain critical value. Therefore, lower bounds for the stress components may be useful for indicating a dangerous state of stress. To this end certain interesting results were found by SIGNORINI [1, 6, 7], while several other inequalities have been established recently (GRIOLI [3, 4, 9, 10]; BRESSAN [3]). The results given in this chapter concern bodies only slightly deformed, so that the classical linear theory is sufficient. However, some of these results remain valid for arbitrary continuous media, but most are valid only for homogeneous, anisotropic, elastic bodies whose elastic potential energy is a positive-definite quadratic form in the stress components.

§ 2. Integral Properties

I begin by presenting some integral properties of the stress which are fundamental for many inequalities in this chapter and even for establishing a method of integration for the linear boundary-value problem of elastic equilibrium (GRIOLI [5]).

On the supposition that the body is only slightly deformed, so that the linear theory is sufficient, the basic equations of statics are those given by the first of the successive systems (4.17). For brevity, I omit the index (1) and suppose that C^* is an unstressed state of equilibrium. Then the basic equations are

$$Y'^s_{rs} = k^* F_r \qquad \text{(in } C^*\text{)} , \qquad (6.1)$$

$$Y_{rs} N^s_* = f^*_r \qquad \text{(on } \Sigma^*\text{)} . \qquad (6.2)$$

Let

$$b^{(r)}_{\eta \tau \lambda} = -\frac{1}{C^*} \left\{ \int_{C^*} y^\eta_1 y^\tau_2 y^\lambda_3 F_r dC^* + \int_{\Sigma^*} y^\eta_1 y^\tau_2 y^\lambda_3 f^*_r d\Sigma^* \right\} , \qquad (6.3)$$

where the numbers η, τ, λ are non-negative integers. Put

$$\eta + \tau + \lambda = n \qquad (n = 0, 1, \ldots) . \qquad (6.4)$$

The quantities $b^{(r)}_{\eta \tau \lambda}$ are the *astatic* coordinates if $n = 1$, the hyperastatic ones if $n = 2$ (SIGNORINI [1]). It is natural to call the $b^{(r)}_{\eta \tau \lambda}$ the *n-astatic* coordinates of the external forces, for any integer n.

The fundamental equations of statics, which assert that the resultant and the resultant moment of the external forces vanish, are expressed in the forms

$$\left.\begin{array}{l} b^{(r)}_{000} = 0 , \\[4pt] b^{(2)}_{100} - b^{(1)}_{010} = 0 , \quad b^{(3)}_{100} - b^{(1)}_{001} = 0 , \quad b^{(3)}_{010} - b^{(2)}_{001} = 0 . \end{array}\right\} \qquad (6.5)$$

Henceforth a bar above a function symbol shall denote its mean value in C^*.

Multiplying (6.1) by $y^\eta_1 y^\tau_2 y^\lambda_3 dC^*$, integrating over C^*, then adding (6.2) multiplied by $y^\eta_1 y^\tau_2 y^\lambda_3 d\Sigma^*$ and integrated over Σ^*, one shows easily that

$$\eta \,\overline{Y_{r1} y^{\eta-1}_1 y^\tau_2 y^\lambda_3} + \tau \,\overline{Y_{r2} y^\eta_1 y^{\tau-1}_2 y^\lambda_3} + \lambda \,\overline{Y_{r3} y^\eta_1 y^\tau_2 y^{\lambda-1}_3} = b^{(r)}_{\eta \tau \lambda} . \qquad (6.6)$$

It is not difficult to see that if the Y_{rs} have singularities owing to concentrated forces, (6.6) remain valid if such forces are properly included in the expressions for $b^{(r)}_{\eta \tau \lambda}$.

Specifically, if concentrated forces or forces applied along lines are present, the corresponding terms are to be added to the second members of (6.3).

For $n = 1, 2$ equations (6.6), known for some time (SIGNORINI [1]), give the mean values in C^* of Y_{rs} and of the products $Y_{rs} y_t$. However, when $n \geqslant 3$, the system is indeterminate, as may easily be foreseen, there being in fact $\frac{3}{2}(n+1)(n+2)$ quantities $b^{(r)}_{\eta \tau \lambda}$, while the number of mean values occurring in (6.6) is $3n(n+1)$. Nevertheless for applications it is interesting that fifteen of the mean values occurring in (6.6) are determinate for any n. Indeed, from (6.6) follow the nine equations

$$n\,\overline{Y_{r1}\,y_1^{n-1}}=b_{n\,00}^{(r)}, \qquad n\,\overline{Y_{r2}\,y_2^{n-1}}=b_{0n0}^{(r)}, \qquad n\,\overline{Y_{r3}\,y_3^{n-1}}=b_{00n}^{(r)} \tag{6.7}$$

and six of the form

$$n\,\overline{Y_{11}\,y_1^{n-1}\,y_2}=b_{n10}^{(1)}-\frac{b_{n+1\,00}^{(2)}}{n+1}, \quad \text{etc.} \tag{6.8}$$

Let P_t be an arbitrary polynomial in y_r. Plainly the mean value $\overline{Y_{rs}\,P_t}$ is a linear combination of some $\overline{Y_{rs}\,y_1^\eta\,y_2^\tau\,y_3^\lambda}$. By saying that $\overline{Y_{rs}\,P_t}$ is of class M I mean that those $\overline{Y_{rs}\,y_1^\eta\,y_2^\tau\,y_3^\lambda}$ satisfy (6.6). Plainly it is possible to construct polynomials P_t, of arbitrarily high degree, such that $\overline{Y_{rs}\,P_t}$ is a uniquely determined linear combination of the $b_{\eta\tau\lambda}^{(r)}$, occurring in (6.7), (6.8). In this case, I shall say that $\overline{Y_{rs}\,P_t}$ is of class M'.

§ 3. A basic inequality

Let $Q_0, Q_1, \ldots Q_m$, be $m+1$ functions of y_1, y_2, y_3, defined and orthogonal in C^*. If $q_{rs}=q_{sr}$ are the constant coefficients of an arbitrary quadratic form in six variables, positive-definite or at least positive semi-definite, let

$$\Psi=\sum_{r,s=1}^{6}\int_{C^*} q_{rs}\,(Y_r-\sum_{t=0}^{m}\gamma_{rt}Q_t)\,(Y_s-\sum_{t=0}^{m}\gamma_{st}Q_t)\,dC^*, \tag{6.9}$$

$$\mu_t^2=\frac{1}{C^*}\int_{C^*} Q_t^2\,dC^* \qquad (t=0,1\ldots,m), \tag{6.10}$$

where $\gamma_{rt}\ (r=1,2\ldots,6;\ t=0,1\ldots,m)$ are arbitrary constants. From (6.9), (6.10) it follows that

$$\Psi=\sum_{r,s=1}^{6}\int_{C^*} q_{rs}Y_r\,Y_s\,dC^*+C^*\sum_{r,s=1}^{6}\sum_{t=0}^{m} q_{rs}\,\gamma_{st}\,(\mu_t^2\gamma_{rt}-2\,\overline{Y_r\,Q_t})\geqslant 0. \tag{6.11}$$

If one assumes that

$$\gamma_{rt}=\frac{\overline{Y_r\,Q_t}}{\mu_t^2} \qquad (r=1,2\ldots,6;\ t=0,1\ldots,m), \tag{6.12}$$

from (6.11) it follows that

$$\sum_{r,s=1}^{6}\int_{C^*} q_{rs}\,Y_r\,Y_s\,dC^*\geqslant C^*\sum_{t=0}^{m}\sum_{r,s=1}^{6} q_{rs}\,\frac{\overline{Y_r\,Q_t}\cdot\overline{Y_s\,Q_t}}{\mu_t^2}. \tag{6.13}$$

It is to be observed that if the quadratic form defined by q_{rs} is actually positive-definite, then the γ_{rs}, which are expressed by (6.12), make Ψ a minimum, since (6.12) is a necessary consequence of the equations

$$\sum_{r=1}^{6} q_{rs} \left(\mu_t^2 \gamma_{rt} - \overline{Y_r Q_t} \right) = 0 \qquad (s = 1, 2 \ldots, 6) , \qquad (6.14)$$

the determinant of q_{rs} being different from zero.

Further, in this case it is clear that the sign of equality is valid in (6.13) if and only if

$$Y_r = \sum_{t=0}^{m} \frac{\overline{Y_r Q_t}}{\mu_t^2} Q_t \qquad (r = 1, 2 \ldots, 6) . \qquad (6.15)$$

Therefore, *among all states of stress corresponding to the same* $\overline{Y_r Q_t}$ *($r = 1 \ldots 6$; $t = 0, 1 \ldots, m$), the stress that minimizes, the mean value of any constant positive-definite quadratic form in the stress components is given by* (6.15).

(6.13) is of interest when one is in a position to give relations between $\overline{Y_r Q_t}$ and the external forces.

In particular, that happens when the functions Q_t are polynomials $P_0, P_1 \ldots, P_m$ orthogonal in C^*, since in this case one may use (6.7), (6.8). If $Q_t \equiv P_t$ ($t = 0, 1 \ldots, m$), (6.13) becomes

$$\sum_{r,s=1}^{6} \int_{C^*} q_{rs} Y_r Y_s \, dC^* \geqslant C^* \sum_{t=0}^{m} \sum_{r,s=1}^{6} q_{rs} \frac{\overline{Y_r P_t} \cdot \overline{Y_s P_t}}{\varrho_t^2} , \qquad (6.16)$$

where

$$\varrho_t^2 = \frac{1}{C^*} \int_{C^*} P_t^2 \, dC^* \qquad (t = 0, 1 \ldots, m) . \qquad (6.17)$$

If the origin of the frame of reference is taken at the center of mass of C^* and the axes are taken as the principal axes of inertia, when $m = 3$ and $P_0 \equiv 1, P_i \equiv y_i$ ($i = 1, 2, 3$), the inequality (6.16), in view of (6.7), (6.8), coincides with the one established by SIGNORINI [1].

For general m, the right-hand side of (6.16) is a known function of $b_{\eta \tau \lambda}^{(r)}$ if the polynomials P_t are such that $\overline{Y_r P_t}$ belongs to class M'. Otherwise the second member of (6.16), which in any case is a quadratic form in $\overline{Y_{rs} y_1^\eta y_2^\tau y_3^\lambda}$, shares in the indeterminateness of the means as expressed by equations (6.6). In the indeterminate case, one may derive from (6.16) an inequality with known second member by identifying $\overline{Y_{rs} y_1^\eta y_2^\tau y_3^\lambda}$ with the values that minimize it, taking account of equations (6.6). If constraints are present, a part of the surface forces is generally unknown, and some of the $b_{\eta \tau \lambda}^{(r)}$ are sums of two terms, one of which, $c_{\eta \tau \lambda}^{(r)}$, is unknown. Then the minimum of the second member of (6.16) is to be determined in the class of all possible $c_{\eta\tau\lambda}^{(r)}$ compatible with the constraints and with the basic equations of statics [i.e. with (6.5)].

Although some of the results obtained are valid for arbitrary functions Q_t if equalities analogous to (6.6) have been established for them, hereafter I shall refer only to orthogonal polynomials P_0, P_1, ..., P_m in C^*, since the integral properties have been defined with reference to them.

§ 4. A few consequences of inequality (6.16)

If special values are given to q_{rs}, from (6.16) some interesting inequalities follow. For example, supposing every q_{rs} to be zero except one with equal indices, one has[1]

$$|Y_r|_{\max} > \sqrt{\sum_{t=0}^{m} \frac{\overline{Y_r P_t}^2}{\varrho_t^2}} \qquad (r = 1, 2 \ldots, 6) . \qquad (6.18)$$

If the external forces are all known, lower bounds for the right-hand member of (6.18) may be gotten by using an arbitrary number of known functions without having to apply the minimizing procedure of the previous section. It is sufficient, for example, to consider the sequence

$$1, y_1, y_2, y_3; y_p^2, y_p^3, \ldots \qquad (6.19)$$

and to suppose that P_t is that linear combination of the first $t + 1$ elements of (6.19) which is orthogonal to the first t elements. In this case, by (6.7) it is evident that the second member of (6.18) becomes known if in the sequence (6.19) one identifies p with r when $r = 1, 2, 3$ and p with 2 or 3, with 1 or 3, or with 1 or 2, when $r = 4, 5$ and 6, respectively. Further, if $r = 1, 2, 3$ in (6.18), one may associate monomials of the type $y_r^{n-2} y_s$ with those of (6.19) [considered for $p = r$] and so increase the second member of (6.18), recalling (6.8).

If $\sum_{r,s=1}^{6} q_{rs} Y_r Y_s$ is taken in turn as the semidefinite quadratic forms giving the square of the normal and tangential stresses acting on a plane with unit normal v, (6.16) gives lower bounds for their magnitudes. In any case v may be determined in such manner as to make the second member of (6.16) a maximum.

By identifying $\sum_{m,s=1}^{6} q_{rs} Y_r Y_s$ with the expressions of the square of ε_{pq}, (6.16) gives lower bounds for $|\varepsilon_{pq}|_{\max}$, etc. It is interesting to observe that a necessary condition of elasticity follows from (6.16), for homogeneous bodies: *A necessary condition in order that the von Mises condition, which*

---

[1] In essence, (6.18) is equivalent to Bessel's inequality. For some inequalities for the stress in curvilinear coordinates see BRESSAN [3].

asserts that plastic yield has not set in, be satisfied is

$$\sum_{t=0}^{m} \frac{1}{\varrho_t^2} \sum_{r=1}^{3} [(\overline{Y_{rr}P_t} - \overline{Y_{r+1\,r+1}\,P_t})^2 + 6\,\overline{Y_{r\,r+1}\,P_t^2}] < b^2 , \qquad (6.20)$$

where b^2 is a constant coefficient characterizing the elastic limit. In effect, (6.20) imposes a limitation on the external forces.

§ 5. A new general inequality

1. I shall now derive certain new inequalities for the stress in a body in equilibrium, valid whatever be its constitution.

Let β_s $(s = 1, 2, \ldots, 6)$ and a_t $(t = 0, 1, \ldots, m)$ be arbitrary constant coefficients. Plainly

$$\int_{C^*} \sum_{s=1}^{6} \beta_s Y_s \sum_{t=0}^{m} a_t P_t \, dC^* = C^* \sum_{s=1}^{6} \beta_s \sum_{t=0}^{m} a_t \overline{Y_s P_t} . \qquad (6.21)$$

From (6.21) one easily deduces[1] that

$$\Big|\sum_{s=1}^{3} \beta_s Y_s\Big|_{\max} > \frac{C^*\Big|\sum_{s=1}^{6}\beta_s\sum_{t=0}^{m} a_t \overline{Y_s P_t}\Big|}{\int_{C^*}\Big|\sum_{t=0}^{m} a_t P_t\Big|\,dC^*} . \qquad (6.22)$$

In writing (6.22) I have omitted the sign of equality which, plainly, may occur in exceptional cases. In considering (6.22), we are to bear in mind (6.6). The second member of (6.22) is a known function of $b_{\eta\tau\lambda}^{(r)}$ if $\overline{Y_s P_t}$ is of class M'. The calculation of the right-hand side of (6.22) is more burdensome than that needed for the corresponding inequality which one may deduce from (6.16) by choice of particular q_{rs}, but inequality (6.22) is certainly more useful if a_t is chosen suitably.

In fact, supposing that

$$\sum_{r,s=1}^{6} q_{rs} Y_r Y_s = \Big[\sum_{s=1}^{6} \beta_s Y_s\Big]^2 , \qquad (6.23)$$

from (6.16) we derive the inequality

$$\Big|\sum_{s=1}^{6} \beta_s Y_s\Big|_{\max} > \sqrt{\sum_{t=0}^{m} \frac{1}{\varrho_t^2} \Big[\sum_{s=1}^{6} \beta_s \overline{Y_s P_t}\Big]^2} . \qquad (6.24)$$

Instead, if one supposes that

$$a_t = a_t' = \sum_{s=1}^{6} \beta_s \frac{\overline{Y_s P_t}}{\varrho_t^2} , \qquad (6.25)$$

---

[1] If all but one of the β_s are zero, see GRIOLI [9].

the second member of (6.22), which I shall denote by η, is greater than that of (6.24). In fact

$$\eta = \frac{C* \sum\limits_{t=0}^{m} \frac{1}{\varrho_t^2} [\sum\limits_{s=1}^{6} \beta_s \overline{Y_s P_t}]^2}{\int\limits_{\mathring{C}*} \left| \sum\limits_{t=0}^{m} \frac{1}{\varrho_{t|}^2} \sum\limits_{s=1}^{6} \beta_s \overline{Y_s P_t} \cdot P_t \right| dC*} , \tag{6.26}$$

from which, by the orthogonality of the P_t and by (6.17), it follows that

$$\eta^2 > C* \frac{\left[\sum\limits_{t=0}^{m} \frac{1}{\varrho_t^2} (\sum\limits_{s=1}^{6} \beta_s \overline{Y_s P_t})^2 \right]^2}{\int\limits_{\mathring{C}*} \left[\sum\limits_{t=0}^{m} \frac{1}{\varrho_t^2} \sum\limits_{s=1}^{6} \beta_s \overline{Y_s P_t} \cdot P_t \right]^2 dC*} = \sum\limits_{t=0}^{m} \frac{1}{\varrho_t^2} [\sum\limits_{s=1}^{6} \beta_s \overline{Y_s P_t}]^2 . \tag{6.27}$$

If all the β_s except one are taken as zero, from (6.22) it follows that

$$|Y_r|_{\max} > C* \frac{\left| \sum\limits_{t=0}^{m} a_t \overline{Y_r P_t} \right|}{\int\limits_{\mathring{C}*} \left| \sum\limits_{t=0}^{m} a_t P_t \right| dC*} , \tag{6.28}$$

which, when a_t is expressed by means of (6.25), becomes

$$|Y_r|_{\max} > C* \frac{\sum\limits_{t=0}^{m} \frac{\overline{Y_r P_t^2}}{\varrho_t^2}}{\int\limits_{\mathring{C}*} \left| \sum\limits_{t=0}^{m} \frac{\overline{Y_r P_t}}{\varrho_t^2} P_t \right| dC*} , \tag{6.29}$$

more useful than (6.18).

2. Let $\boldsymbol{v} \equiv (v_1, v_2, v_3)$ be an arbitrary unit vector, and let

$$\beta_s = v_s^2 , \qquad \beta_{s+3} = 2 v_{s+1} v_{s+2} \qquad (s = 1, 2, 3) . \tag{6.30}$$

The normal component Φ_{vn} of the stress in the direction of $\boldsymbol{v}$ is simply $\sum\limits_{s=1}^{6} \beta_s Y_s$. Then (6.22) gives

$$|\Phi_{vn}|_{\max} > C* \frac{\left| \sum\limits_{t=0}^{m} a_t \overline{\Phi_{vn} P_t} \right|}{\int\limits_{\mathring{C}*} \left| \sum\limits_{t=0}^{m} a_t P_t \right| dC*} , \tag{6.31}$$

of value at least in the case when a_t is expressed by (6.25). Plainly inequalities (6.22), (6.28), (6.31) give the most useful results when one chooses a_t so as to maximize the second members of those inequalities.

§ 6. Application of (6.28) to the equilibrium of cylindrical bodies

Inequality (6.28) gives interesting results if C^* is a cylinder of section A^*, even if one considers only polynomials P_t of the first degree (GRIOLI[9]).

Let us suppose that the reference trihedral $O^* y_1 y_2 y_3$ has its origin at the center of mass and its axes coinciding with the principal axes of inertia, y_1 being parallel to the generators of C^*. I shall deduce from (6.28) an inequality for the normal stress Y_{11}, the evaluation of which is often of greatest interest in the applications.

Let $C'(y_1)$ be the portion of C^* bounded by the base on which $y_1 < 0$ and by the typical section $y_1 = \text{const.}$, and let $R'_i(y_1)$, $M'_i(y_1)$ be the components in the coordinate system $O^* y_1 y_2 y_3$ of the resultant force and moment, with respect to the center of mass of the section $y_1 = \text{const.}$, of the external forces applied upon $C'(y_1)$. If R_i, M_i are the mean values of $R'_i(y_1)$, $M'_i(y_1)$ when y_1 varies between the two bases of the body

$$\overline{Y_{11}} = \frac{R_1}{A^*}, \qquad \overline{Y_{11} y_2} = -\frac{M_3}{A^*}, \qquad \overline{Y_{11} y_3} = \frac{M_2}{A^*}. \qquad (6.32)$$

When $m = 3$ and

$$P_0 = 1, \qquad P_1 = 0, \qquad P_2 = y_2, \qquad P_3 = y_3, \qquad (6.33)$$

then $\overline{Y_{11} P_t}$ $(t = 0, 2, 3)$ coincide with the right-hand members of (6.32), and (6.28) for $r = 1$ becomes

$$|Y_{11}|_{\max} > \frac{a_0 R_1 - a_2 M_3 + a_3 M_2}{\int_{A^*} |a_0 + a_2 y_2 + a_3 y_3| \, dA^*}. \qquad (6.34)$$

The constants a_0, a_2, a_3 should be determined so as to maximize the second member of (6.34).

Nevertheless, one may avoid finding such values of a_0, a_2, a_3, which may be laborious because of the shape of the section, instead identifying the a_t with the values a'_t expressed by (6.25), for $\beta_1 = 1$, $\beta_s = 0$ $(s = 2, 3, \ldots, 6)$. It follows that

$$|Y_{11}|_{\max} > z \qquad (6.35)$$

with

$$z = \frac{R_1^2 + \dfrac{M_3^2}{\varrho_2^2} + \dfrac{M_2^2}{\varrho_3^2}}{\displaystyle\int_{A^*} \left| R_1 - \dfrac{M_3}{\varrho_2^2} y_2 + \dfrac{M_2}{\varrho_3^2} y_3 \right| dA^*}, \qquad (6.36)$$

where

$$\varrho_i^2 = \frac{1}{A^*} \int_{A^*} y_i^2 \, dA^* . \tag{6.37}$$

If M_2, M_3 are not both zero, one may give a revealing form to z. Let us suppose, as is certainly permitted, that $R_1 \geqslant 0$, and let us call τ the denominator of the second member of (3.36) and r the straight line having[1] the equation

$$R_1 - \frac{M_3}{\varrho_2^2} y_2 + \frac{M_2}{\varrho_3^2} y_3 = 0 \tag{6.38}$$

in the plane π of A^*.
   One has

$$\tau = \int_{A_1} \left[R_1 - \frac{M_3}{\varrho_2^2} y_2 + \frac{M_2}{\varrho_3^2} y_3 \right] dA$$

$$- \int_{A^*-A_1} \left[R_1 - \frac{M_3}{\varrho_2^2} y_2 + \frac{M_2}{\varrho_3^2} y_3 \right] dA^* , \tag{6.39}$$

where A_1 is the portion of A^* which belongs to the half plane of π where the first member of (6.38) is positive, if r crosses A^*, or is all of A^* in the contrary case[2]. Bearing in mind that the co-ordinate axes are the principal axes of inertia, from (6.39) we have

$$\tau = (2 A_1 - A^*) R_1 + 2 A_1 \left[\frac{M_2}{\varrho_3^2} y_3^{(1)} - \frac{M_3}{\varrho_2^2} y_2^{(1)} \right], \tag{6.40}$$

where $y_2^{(1)}$, $y_3^{(1)}$ denote the coordinates y_2, y_3 of the center of mass G_1 of A_1. Let M^* be the vector whose components are $0, \dfrac{M_2}{\varrho_3^2}, \dfrac{M_3}{\varrho_2^2}$. The part of equation (6.40) which depends upon M_2, M_3 is $|M^* \times OG_1|$; it coincides[3] with $M^* \delta_1$, if δ_1 denotes the distance of G_1 from the parallel to M^* through the center of mass G of A^*. Therefore,

$$\tau = (2 A_1 - A^*) R_1 + 2 A_1 M^* \delta_1 . \tag{6.41}$$

---

[1] In the following considerations and in those of §§ 7, 8, A^* may be identified with the section $y_1 = 0$.
[2] The center of mass of A^* belongs to A_1, since $R_1 \geqslant 0$.
[3] The straight line having the equation

$$\frac{M_2}{\varrho_3^2} y_3 - \frac{M_3}{\varrho_2^2} y_2 = 0 \tag{*}$$

pivides the section A^* in two parts, in one of which, A_1', the left-hand side of (*) is positive. Plainly, for $R_1 > 0$, A_1' is a portion of A_1 and contains G_1. Hence it follows that $\dfrac{M_2}{\varrho_3^2} y_3^{(1)} - \dfrac{M_3}{\varrho_2^2} y_2^{(1)} \geqslant 0$.

The distance of r from G is $\frac{R_1}{M^*}$; hence

$$\delta_1 = \delta - \frac{R_1}{M^*},\tag{6.42}$$

where δ is the distance of G_1 from r.

Finally, then, one has

$$\tau = 2\,A_1\,M^*\,\delta - A^*\,R_1 > 0.\tag{6.43}$$

Correspondingly, (6.35) becomes [see (6.36)]

$$|Y_{11}|_{max} > \frac{R_1^2 + M \cdot M^*}{2A_1\,M^*\,\delta - A^*\,R_1}.\tag{6.44}$$

If r does not cross A^*, then $A_1 = A^*$, $\delta = \frac{R_1}{M^*}$, and (6.44) becomes

$$|Y_{11}|_{max} > \frac{R_1^2 + M \cdot M^*}{A^*\,M^*\,\delta}.\tag{6.45}$$

In the particular case of astatic external forces, $R_1 = 0$ and (6.44) becomes

$$|Y_{11}|_{max} > \frac{M \cdot M^*}{2A_1\,M^*\,\delta},\tag{6.46}$$

where δ denotes the distance of G_1 from the straight line parallel to M^* for G. It is easy to see that the point Q^* having coordinates $\frac{M_3}{R_1}$, $-\frac{M_2}{R_1}$ is the anti-pole of the straight line r with respect to the principal ellipse of A^* centered at the center of mass. Then r does not cross the section A^* if, and only if, Q^* is internal to the central kernel. One deduces that the first or the second of the inequalities (6.44), (6.45) holds according as Q^* is internal or external to the central kernel of A^*. In a certain sense r and Q^* correspond to the *neutral axis* and to the *center of loading*, according to their definitions in Saint-Venant's formulation of the problem of simultaneous extension and bending. But in this case r and Q^* become just the *neutral axis* and the *center of loading*, while R_1, M_2, M_3 coincide with the components with respect to the axes of the reference frame of the resultant force and moment with respect to the center of mass of C^*, of the external forces applied on the base where $y_1 > 0$.

The validity of inequalities (6.34), (6.35) may be extended, by easy modifications, to the case of non-cylindrical bodies having principal axes of inertia parallel to the central ones and having their origin at any point on the axis of y_1. In fact, it is easy to see that (6.32) remains valid if one takes $A^* = \frac{C^*}{l}$ with $l = y_{1\,max} - y_{1\,min}$, and that from (6.25), (6.28) follow the inequalities obtained from (6.34), (6.35), (6.36) by multiplying the second members of (6.34), (6.35) by l and then substituting for the integrals analogous ones with C^* as the domain of integration. Naturally, ϱ_i^2 are to be replaced by (6.17).

§ 7. On the best a_t in a case similar to that of uniform bending

Considering the cylindrical body described in the last section, I shall deal with a particularly interesting case which may be regarded as the counterpart of the problem of pure bending if the external forces act only on the bases of the cylinder.

Inequality (6.28) for $r = 1$ and with reference only to polynomials $P_2 = y_2$, $P_3 = y_3$ becomes

$$|Y_{11}|_{\max} > \frac{|a_3 M_2 - a_2 M_3|}{\int\limits_{A^*} |a_2 y_2 + a_3 y_3| \, dA^*} .\qquad(6.47)$$

As is plainly allowed, let us put

$$a_2 = -\sin \alpha , \qquad a_3 = \cos \alpha . \qquad(6.48)$$

Consequently (6.47) becomes

$$|Y_{11}|_{\max} > z(\alpha) \qquad(6.49)$$

with

$$z(\alpha) = \frac{|M_2 \cos \alpha + M_3 \sin \alpha|}{\nu} \qquad(6.50)$$

and

$$\nu = 2 \int\limits_{A_1} (y_3 \cos \alpha - y_2 \sin \alpha) \, dA^* , \qquad(6.51)$$

where A_1 is the part of the cross-section where

$$y_3 \cos \alpha - y_2 \sin \alpha > 0 . \qquad(6.52)$$

Calling r' the line $y_3 \cos \alpha - y_2 \sin \alpha = 0$, if we attach a unit vector to r' we may say, specifically, that α is the angle between r' and positive semi-axis of y_2, supposing that $\alpha \geqslant 0$ when the positive half-line r' belongs to the half-plane $y_3 = 0$ and that $\alpha \leqslant 0$ in the opposite case.

If an assigned value of α is given an infinitesimal increment $\Delta\alpha$, an increment ΔA_1 of A_1 corresponds to it. The consequent increment $\Delta\nu$ of ν is

$$\Delta\nu = 2 \int\limits_{A_1 + \Delta A_1} [y_3 \cos(\alpha + \Delta\alpha) - y_2 \sin(\alpha + \Delta\alpha)] \, dA^*$$

$$- 2 \int\limits_{A_1} (y_3 \cos \alpha - y_2 \sin \alpha) \, dA^* , \qquad(6.53)$$

which, neglecting infinitesimals of order greater than $\Delta\alpha$, becomes

$$\Delta\nu = -2 \Delta\alpha \int\limits_{A_1} (y_2 \cos \alpha + y_3 \sin \alpha) \, dA^* . \qquad(6.54)$$

It follows that

$$\frac{dv}{d\alpha} = -2 A_1 (y_2^{(1)} \cos \alpha + y_3^{(1)} \sin \alpha),\tag{6.55}$$

where, as in the last section, $y_2^{(1)}$, $y_3^{(1)}$ are the coordinates of index 2, 3 of the center of mass G_1 of A_1. From (6.50), (6.55) it follows that

$$\frac{dz}{d\alpha} = \pm 2 A_1 \frac{M_2 y_2^{(1)} + M_3 y_3^{(1)}}{v^i},\tag{6.56}$$

where the upper or lower signs are to be taken according as $M_2 \cos \alpha + M_3 \sin \alpha \lessgtr 0$.

For simplicity, I suppose that any straight line passing through the center of mass of the cross-section of the cylindrical body and lying in its plane crosses the boundary at two points only. Then in a polar co-ordinate system in the section $A*$ with the pole at its center of mass and the polar axis parallel to the axis of y_2 the polar equation of the boundary has the form

$$\varrho = R(\varphi) > 0.\tag{6.57}$$

Put

$$\lambda(\varphi) = M_2 \cos \varphi + M_3 \sin \varphi;\tag{6.58}$$

it follows that

$$A_1 (M_2 y_2^{(1)} + M_3 y_3^{(1)}) = \int_{A_1} \lambda(\varphi) \varrho^2 d\varrho \, d\varphi = \frac{1}{3} \int_{\alpha}^{\alpha+\pi} \lambda(\varphi) R^3(\varphi) d\varphi.\tag{6.59}$$

Without essential loss of generality, I shall suppose that $M_2 > 0$, $M_3 > 0$. One easily perceives the existence of values $\alpha*$ of α which make $\frac{dz}{d\alpha}$ zero. For the purpose it is sufficient to note that, if $\varphi*$ is the unique value of φ lying between $\frac{\pi}{2}$ and π which makes $\lambda(\varphi)$ vanish, then the integral of the third member of (6.59) operates on a function which is always non-positive when $\alpha = \varphi*$, always non-negative when $\alpha = \varphi* - \pi$. Therefore, the sign of $\frac{dz}{d\alpha}$ changes when α varies between $\varphi* - \pi$ and $\varphi*$. From (6.56), (6.58), (5.69) we deduce that for any value of α which makes $\frac{dz}{d\alpha}$ zero, we have

$$\frac{d^2 z}{d\alpha^2} = \mp \frac{M_2 \cos \alpha* + M_3 \sin \alpha*}{2 v^2},\tag{6.60}$$

where the upper or lower signs are to be taken according as the numerator of the fraction in the second member is positive or negative. This means that $\frac{d^2 z}{d\alpha^2}$ is negative for any $\alpha*$, and $z(\alpha)$ has a maximum for $\alpha = \alpha*$.

§ 8. Rectangular prism. A geometrical interpretation of the results

1. Now I suppose that the section A^* is a rectangle, and I denote by $2b_2$, $2b_3$ the lengths of the sides and by $q = \dfrac{b_3}{b_2} < 1$ their quotient.

For evident reasons of symmetry one may suppose that $0 < \alpha < \pi$. Briefly, it turns out that

$$A_1(M_2 y_2^{(1)} + M_3 y_3^{(1)}) \begin{cases} = \dfrac{b_2}{3}\left[(3\,b_3^2 - b_2^2\,\mathrm{tg}\,^2\alpha)\,M_3 - 2\,b_2^2 M_2\,\mathrm{tg}\,\alpha\right] \\ \qquad\qquad (\text{for } 0 < |\mathrm{tg}\,\alpha| < q)\,, \\[2mm] = \dfrac{b_3}{3\,\mathrm{tg}\,^2\alpha}\left[(b_3^2 - 3\,b_2^2\,\mathrm{tg}\,^2\alpha)\,M_2 + 2\,b_3^2 M_3\,\mathrm{tg}\,\alpha\right] \\ \qquad\qquad (\text{for } |\mathrm{tg}\,\alpha| > q)\,. \qquad\qquad (6.61) \end{cases}$$

Put

$$\eta = \frac{M_2}{M_3} > 0; \qquad\qquad (6.62)$$

then it is evident that in the interval from 0 to q the equation

$$\mathrm{tg}^2\,\alpha + 2\eta\,\mathrm{tg}\,\alpha - 3q^2 = 0 \qquad\qquad (6.63)$$

admits a unique solution, expressed by

$$\mathrm{tg}\,\alpha^* = -\eta + \sqrt{\eta^2 + 3q^2}\,, \qquad\qquad (6.64)$$

if and only if $\eta > q$, while in the interval from q to ∞ the equation

$$\mathrm{tg}^2\,\alpha - \frac{2}{3}\frac{q^2}{\eta}\,\mathrm{tg}\,\alpha - \frac{q^2}{3} = 0 \qquad\qquad (6.65)$$

admits a unique solution, namely,

$$\mathrm{tg}\,\alpha^* = \frac{q^2}{3\eta}\left[1 + \sqrt{1 + \frac{3\eta^2}{q^2}}\right], \qquad\qquad (6.66)$$

if and only if $\eta < q$.

Instead, if $\eta > 0$, (6.63) admits no solution in the interval from 0 to $-q$, and (6.65) does not admit any in the interval from $-q$ to $-\infty$. By what was said above one deduces that $z(\alpha)$ assumes its maximum value for $\alpha = \alpha^*$, if α^* is given by (6.64), or (6.66), according as $\eta > q$ or $\eta < q$. Correspondingly,

$$z(\alpha^*) \begin{cases} = \dfrac{M_3[\eta + \sqrt{\eta^2 + 3q^2}]}{4b_2 b_3^2} & (\text{for } \eta > q)\,, \\[4mm] = \dfrac{M_3\left[1 + \sqrt{1 + \dfrac{3\eta^2}{q^2}}\right]}{4b_2^2 b_3} & (\text{for } 0 < \eta < q)\,. \end{cases} \qquad (6.67)$$

It is easily seen that for $\eta = q$ the expressions (6.64), (6.66) for tg α^* become equal. Also the expressions (6.67) for $z(\alpha^*)$ become equal.

2. If $|Y_{11}|_{max}$ must not exceed a certain value $k > 0$, the inequality (6.49) gives a necessary condition of safety, which may have an interesting geometrical interpretation because of (6.67). I begin by observing that this condition implies that

$$k > z(\alpha^*) ; \tag{6.68}$$

if we put

$$\zeta_2 = \frac{M_3}{A^* k} , \qquad \zeta_3 = \frac{M_2}{A^* k} , \tag{6.69}$$

and bear in mind (6.62), (6.67), this result leads to the inequalities

$$\zeta_3^2 + \frac{2 b_2 q^2}{3} \zeta_2 - \frac{b_3^2}{3} < 0 , \qquad \text{for } 0 \leqslant \eta = \frac{\zeta_3}{\zeta_2} \leqslant q , \tag{6.70}$$

$$\zeta_2^2 + \frac{2 b_3}{3 q^2} \zeta_3 - \frac{b_2^2}{3} < 0 , \qquad \text{for } \eta = \frac{\zeta_3}{\zeta_2} \geqslant q . \tag{6.71}$$

Let $\mathfrak{P}_1^*$ be the parabola given by the equation

$$y_3^2 + \frac{2 b_2 q^2}{3} y_2 - \frac{b_3^2}{3} = 0 , \tag{6.72}$$

and let B_1^* be the portion of the half-plane $y_3 \geqslant 0$ bounded by the positive semi-axis of y_2, by the straight line whose equation is

$$y_3 - y_2 q = 0 \tag{6.73}$$

and by $\mathfrak{P}_1^*$. Condition (6.68) when $0 \leqslant \eta \leqslant q$ is equivalent to this: the point $P^* \equiv (\zeta_2, \zeta_3)$ is not external to B_1^* and does not belong to $\mathfrak{P}_1^*$.

Instead, if $\mathfrak{P}_2^*$ is the parabola given by the equation

$$y_2^2 + \frac{2 b_3}{2 q^2} y_3 - \frac{b_2}{3} = 0 , \tag{6.74}$$

and if B_2^* is the portion of half-plane $y_3 \geqslant 0$ bounded by the positive semi-axis $y_3 > 0$, by the straight line given by equation (6.73) and by $\mathfrak{P}_2^*$, the condition (6.68) when $\eta \geqslant q$ becomes the following: P^* is not external to B_2^* and does not belong to $\mathfrak{P}_2^*$.

Definitively, observing that $\mathfrak{P}_1^*$ and $\mathfrak{P}_2^*$ cross the straight line given by equation (6.73) at the same point $\left(\frac{b_2}{3}, \frac{b_3}{3}\right)$, one concludes that condition (6.68), when $\eta \geqslant 0$, $M_3 > 0$, is equivalent to the requirement that the point P^* be not external to region $B_1^* + B_2^*$ and not on $\mathfrak{P}_1^*$ or $\mathfrak{P}_2^*$. The results may be extended by symmetry, dropping any hypothesis on the values of M_2, M_3 and supposing η arbitrary. Therefore, if B^* is the portion of the plane π constituted by the region $B_1^* + B_2^*$ and by its three reflections across the axes of y_2, y_3 one has: *a necessary condition that* $|Y_{11}|_{max}$ *be not greater than k is that the point $P^* \equiv (\zeta_2, \zeta_3)$ be internal to B^*.*

It is easy to see that B^* is internal to the rectangular region bounded by the straight lines $y_2 = \pm \frac{b_2}{2}$, $y_3 = \pm \frac{b_3}{2}$. B^* is also internal to the region bounded by the central ellipse of the section A^*, already indicated by SIGNORINI [6] as the region to which P^* must necessarily belong in order that the above condition of safety may be satisfied, independently of the shape of A^*. For example, if P^* lies on the positive semi-axis y_2, it belongs to B^* only if $0 < \zeta_2 < \frac{b_2}{2}$ and $0 < M_3 < A^* k \frac{b_2}{2}$, while the condition that P^* be interior to the central ellipse of the section A^* requires that $0 < \zeta_2 < \frac{b_2}{\sqrt{3}}$ and $0 < M_3 < A^* k \frac{b_2}{\sqrt{3}}$.

§ 9. Inequalities concerning the equilibrium of an arc

One may establish inequalities such as (6.34), (6.35), (6.36) also when C^* has the shape of a pipe with constant section generated by a plane surface moving perpendicularly along a curve L described by its center of mass. In this case it is convenient to refer the body to curvilinear coordinates $y_1\, y_2\, y_3$ such that y_1 is arc length along L while y_2 and y_3 are distances from the centroid, measured along the principal axes of the section $y_1 = $ const. Under this hypothesis Y_{11} means the normal stress at the points of section $y_1 = $ const., and for any value of y_1 we have the equations

$$\int_{A^*} Y_{11} dA^* = R_1'(y_1), \qquad \int_{A^*} Y_{11} y_2 dA^* = -M_3'(y_1),$$
$$\int_{A^*} Y_{11} y_3 dA^* = M_2'(y_1), \qquad\qquad\qquad (6.75)$$

where R_1', M_2', M_3' have meanings analogous to that used in § 6 but with reference to the frame constituted by the perpendicular to the section A^* at its center of mass and by the principal axes of A^*. With a procedure analogous to the one by which (6.34), (6.35) have been derived one finds the same inequalities (6.34), (6.35), with $R_1(y_1)$ replaced by $R_1'(y_1)$, etc. Such inequalities are valid at any section, and it may be useful to find the one where their second members are maximized. The considerations regarding (6.34), (6.35) may be repeated in this case. By integrating both sides of the inequalities, one may avoid having to determine maxima of the second members.

In particular, one has in $C*$

$$|Y_{11}|_{max} > \frac{1}{l} \int_{-\frac{l}{2}}^{\frac{l}{2}} \frac{R_1'^2 + \frac{M_3'^2}{\varrho_2^2} + \frac{M_2'^2}{\varrho_3^2}}{\int_{A*} \left| R_1' - \frac{M_3'}{\varrho_2^2} y_2 + \frac{M_2'}{\varrho_3^2} y_3 \right| dA*} dy_1 \qquad (6.76)$$

where l denotes the length of L.

The comparison between (6.18) and (6.28), which holds also for (6.34) and (6.35), is not valid here, but others may be made.

§ 10. Inequalities regarding the deformation of slightly deformable bodies

1. The inequalities of the previous sections apply to the stress in bodies in equilibrium, whatever be the constitution of the continuous medium. Inequalities based on (6.22) are generally more profitable, but the corresponding calculations more laborious. Instead, inequality (6.16) has broader possibilities and often makes it possible to get an idea of the deformation of the body. Naturally, in this case the body's physical constitution is brought to bear through the elastic coefficients.

Let us identify the coefficients q_{rs} of inequality (6.16) with the m_{rs} in the quadratic form in Y_r which expresses twice the density of the elastic potential energy for homogeneous bodies. Then, by CLAPEYRON's theorem, the left-hand member of (6.16) becomes equal to the work $\delta\mathfrak{L}$ done by the external forces in the isothermal displacement from the unstressed state to the actual state of equilibrium. Therefore

$$\delta\mathfrak{L} \geqslant C* \sum_{i=0}^{m} \frac{1}{\varrho_i^2} \sum_{r,s=1}^{6} m_{rs} \overline{Y_r P_t} \cdot \overline{Y_s P_t}. \qquad (6.77)$$

This inequality gives a lower bound for twice the density of elastic potential energy stored by the body and often makes it possible to get an idea of the deformation itself, if it is possible to express explicitly the characteristic elements of deformation in its first member.

For example, let us consider an elastic body clamped on a portion of its boundary, supported without friction on another portion and subjected to applied external forces acting in a fixed direction, as is the case for heavy bodies. Then the work of the forces due to the constraints is zero, and from (6.77) it follows that

$$|u|_{max} \geqslant \frac{z}{R}, \qquad (6.78)$$

where z denotes the second member of (6.77), u the component of displacement in the direction of the applied forces and R the magnitude of their resultant.

I have verified that when C^* is a bar clamped or supported at the ends, if one considers polynomials P_t up to the third degree, the second member of (6.78) gives practically the maximum deflection according to the usual theory, for certain sets of applied external forces.

2. An interesting inequality may be derived from (6.77) in the following case (GRIOLI [10]). Let us suppose that on a portion σ' of the boundary of C^* the external surface forces $\boldsymbol{f}'$ act in a fixed direction, characterized by the unit vector $\boldsymbol{v}$. Then

$$\boldsymbol{f}' = f'\boldsymbol{v}, \qquad R' = \int_{\sigma'} f'd\sigma' \lessgtr 0. \tag{6.79}$$

If C' is the center[1] of those forces and $\delta C'$ its displacement corresponding to the deformation of the body, the work done by those external forces is

$$\delta \mathfrak{L}' = \boldsymbol{R}' \cdot \delta C'. \tag{6.80}$$

Let us suppose that on another part σ'' of the boundary acts a second set of external surface forces $\boldsymbol{f}''$, also parallel to $\boldsymbol{v}$, which is in equilibrium with the first one.

Then

$$\boldsymbol{f}'' = f''\boldsymbol{v}, \qquad R'' = \int_{\sigma''} f''d\sigma'' = -\int_{\sigma'} f'd\sigma', \tag{6.81}$$

$$\delta \mathfrak{L}'' = \boldsymbol{R}'' \cdot \delta C'' \tag{6.82}$$

where the meaning of $\delta \mathfrak{L}''$, C'' is analogous to that of $\delta \mathfrak{L}'$, C'. If body forces are absent, the work done by all external forces is expressed by

$$\delta \mathfrak{L} = \delta \mathfrak{L}' + \delta \mathfrak{L}'' = \boldsymbol{R}' \cdot (\delta C' - \delta C''). \tag{6.83}$$

Put

$$C'C'' = l\,\boldsymbol{v}; \tag{6.84}$$

it follows that

$$\delta C' - \delta C'' = -\delta l\,\boldsymbol{v} - l\,\boldsymbol{\Psi} \cdot \boldsymbol{v}, \tag{6.85}$$

and then

$$\delta \mathfrak{L} = -\delta l\,\boldsymbol{R}' \cdot \boldsymbol{v} = -\delta l\,R'. \tag{6.86}$$

From (6.77), (6.86) it follows that

$$|\delta l| \geqslant \frac{z}{|R'|}, \tag{6.87}$$

---

[1] The point C', like C'' [see below], need not correspond to any material element. However, even in this case the results are indicative of the deformation of the body.

which gives an indication of the variation of length in the body in the direction of the external forces. By (6.87) it is possible to verify that *the deformations given by Saint-Venant's theory in the problem of simple extension are smaller than those actually occurring.*

Let us suppose that the reference frame is the central one for the region occupied by the body and take $P_0 \equiv 1$, $P_t \equiv y_t$ $(t = 1, 2, 3)$, $P_t \equiv 0$ $(t > 3)$. From (6.7), (6.8), (6.79), (6.81) it follows that

$$
\left.
\begin{aligned}
\overline{Y_{rs}} &= -\frac{\varphi_r}{C^*}\left\{ \int_{\sigma'} f' y_s d\sigma' + \int_{\sigma''} f'' y_s d\sigma'' \right\}, \\[2mm]
\overline{Y_{rs}\,y_t} &= -\frac{1}{2C^*}\left\{ \int_{\sigma'} f'\left[y_r y_t\,\varphi_s + y_s y_t\,\varphi_r - y_r y_s\,\varphi_t \right] d\sigma' + \right. \\[2mm]
&\quad \left. + \int_{\sigma''} f''\left[y_r y_t\,\varphi_s + y_s y_t\,\varphi_r - y_r y_s\,\varphi_t \right] d\sigma'' \right\},
\end{aligned}
\right\} \tag{6.88}
$$

where the φ_r are the direction cosines of v. From (6.88, 1) it follows that

$$
\overline{Y_{rs}} = \frac{R'}{C^*}\, l\, \varphi_r\, \varphi_s . \tag{6.89}
$$

I shall consider two particularly simple cases.

3. *Case I.* Let there be a correspondence between the points P', P'' of σ' and σ'' such that $P'P''$ is parallel to v and the vectors $f'd\sigma'$, $f''d\sigma''$ at corresponding points are equal and opposite. I denote by y'_r the coordinates of P', by y''_r those of P'', and by q_r those of the mid-point P of $P'P''$. Put

$$
P'P'' = a\, v ; \tag{6.90}
$$

then

$$
y'_r = q_r - \frac{a}{2}\,\varphi_r , \qquad y''_r = q_r + \frac{a}{2}\,\varphi_r , \qquad f'd\sigma' + f''d\sigma'' = 0 . \tag{6.91}
$$

Now (6.88, 2) becomes

$$
\overline{Y_{rs}\,y_t} = \frac{\varphi_r\,\varphi_s}{C^*} \int_{\sigma'} f'\, a\, y_t d\sigma' . \tag{6.92}
$$

Put

$$
\eta_t = \frac{\int_{\sigma'} f'\, a\, y_t d\sigma'}{R'l} . \tag{6.93}
$$

Finally, then, from (6.89), (6.92) one has

$$
\overline{Y_{rs}\,y_t} = \frac{R'l\,\eta_t}{C^*}\,\varphi_r\,\varphi_s = \overline{Y_{rs}}\,\eta_t , \tag{6.94}
$$

where, plainly, if a does not depend upon P, the η_t coincide with the coordinates of the mid-point of $C'C''$. From (6.77), (6.87), (6.94) one derives the inequality

$$|\delta l| \geqslant \frac{C^*}{|R'|} \sum_{r, s=1}^{6} m_{rs} \, \overline{Y}_r \, \overline{Y}_s \left[1 + \sum_{i=1}^{3} \frac{\eta_i^2}{\varrho_i^2}\right]. \tag{6.95}$$

In the particular case of isotropic bodies (6.95) becomes

$$|\delta l| \geqslant \frac{R'l}{EA^*}\left[1 + \sum_{i=1}^{3} \frac{\eta_i^2}{\varrho_i^2}\right], \tag{6.96}$$

where $A^* = \dfrac{C^*}{l}$ and E denotes Young's modulus.

4. *Case II*. Let us suppose that v is directed in the positive sense of the central axis y_1, and suppose there exists a correspondence between σ' and σ'' such that to any point P' of σ' there corresponds a point P'' of σ'' with $P' P''$ parallel to v and with the mid-point of $P' P''$ lying in the plane $y_1 = 0$.

Under this hypothesis, from (6.89) it follows that

$$\overline{Y}_{11} = \frac{R'l}{C^*}, \qquad \overline{Y}_{rs} = 0, \qquad \text{for } r \text{ or } s \neq 1. \tag{6.97}$$

Put $a = |P' P''|$ and

$$\left.\begin{aligned}
\eta_i' &= \frac{1}{l\,R'} \int_{\sigma'} a\,y_i'\, f'\, d\sigma' , \\[2mm]
\eta_i'' &= \frac{1}{l\,R''} \int_{\sigma''} a\,y_i''\, f''\, d\sigma'' \qquad (i = 2, 3) ,
\end{aligned}\right\} \tag{6.98}$$

$$\eta_i^* = \frac{\eta_i' + \eta_i''}{2} . \tag{6.99}$$

From (6.88, 2), (6.97) it follows that

$$\overline{Y_{11} y_t} = \frac{R'\,l}{C^*}\eta_t^* = \overline{Y}_{11}\eta_t^* . \tag{6.100}$$

Since now

$$\overline{Y_{ii} y_1} = 0 \qquad (i = 2,3) , \tag{6.101}$$

from the inequality (6.87), keeping in mind the meaning of z and (6.97), (6.100), for an isotropic body we show that

$$|\delta l| \geqslant \frac{|R'|\,l}{EA^*}\left[1 + \sum_{i=2}^{3} \frac{\eta_i^{*2}}{\varrho_i^2}\right]. \tag{6.102}$$

There is an interesting analogy between the second members of (6.96), (6.102) and the well known expression for the lengthening or shortening given by SAINT-VENANT's theory in the case of simultaneous pressure and bending of a cylindrical body. Indeed, if the body is simply a cylinder subjected to external forces only on its bases [$a = $ const.], then η_t and η_t^* coincide with the coordinates of the mid-point of $C'C''$, and the right-hand members of (6.96), (6.102) express just those magnitudes. This clearly shows that in the theory of simultaneous pressure and bending according to SAINT-VENANT's scheme, one finds smaller deformations than the ones predicted by the exact theory.

For brevity I refrain from presenting inequalities which may be deduced from (6.22) by identification of $\sum_{s=1}^{6} \beta_s Y_s$ with the linear forms which express linear dilatation, shearing, cubical dilatation, etc.

§ 11. On the deformation of a homogeneous elastic shell under pressure

1. Let C^* be a homogeneous elastic body whose boundary consists of two closed surfaces σ_e, σ_i, without points in common. The body is subjected to constant pressures p_e, p_i on the surfaces σ_e, σ_i. I take p_e, p_i as positive if they are pressures, negative in the opposite case.

Further, let V_e, V_i be the volumes of the regions bounded by σ_e, σ_i and y'_r, y''_r the coordinates of the centers of mass of V_e, V_i with respect to the central axes of inertia $\mathfrak{C}^G$ of the body. It is easy to demonstrate that (GRIOLI [3])

$$\overline{Y}_{rs} \begin{cases} = \dfrac{1}{C^*} (V_e\, p_e - V_i\, p_i) & (r = s)\,, \\[2mm] = 0 & (r \neq s)\,, \end{cases} \qquad (6.103)$$

$$\overline{Y_{rs}\, y_t} \begin{cases} = \dfrac{V_e}{C^*}\, y'_r\, (p_e - p_i) = \dfrac{V_i}{C^*}\, y''_r\, (p_e - p_i) & (r = s)\,, \\[2mm] = 0 & (r \neq s)\,. \end{cases} \qquad (6.104)$$

Let us consider in (6.16) only the terms corresponding to $m = 3$ and $P_0 \equiv 1$, $P_t \equiv y_t$. Then, from (6.16), (6.103), (6.104) it follows that

$$\sum_{r,s=1}^{6} \int_{C^*} q_{rs} Y_r Y_s dC^* \geqq \frac{1}{C^*} \sum_{r,s=1}^{3} q_{rs} \Big[(p_e\, V_e - p_i\, V_i)^2 +$$

$$+ V_e^2\, (p_e - p_i)^2 \sum_{i=1}^{3} \frac{y'_r\, y'_s}{\varrho_i^2} \Big]. \qquad (6.105)$$

where, if q_{rs} are the coefficients of a positive-definite quadratic form, the sign of equality is valid if, and only if, $p_e = p_i = p$. It is well known that in this case $Y_{rr} = p$ and $Y_{rr+1} = 0$. Plainly, it is possible to deduce a lower bound for the maximum of $|Y_{rr}|_{max}$ from (6.105). In particular, if $p_i = 0$,

$$|Y_r|_{max} > \frac{V_e |p_e|}{C^*} \sqrt{1 + y_r'^2 \sum_{t=1}^{3} \frac{1}{\varrho_t^2}} =$$

$$= |p_e| \left(1 + \frac{V_i}{C^*}\right) \sqrt{1 + y_r'^2 \sum_{t=1}^{3} \frac{1}{\varrho_t^2}} \quad (r = 1, 2, 3), \quad (6.106)$$

where the sign of equality holds if, and only if, $V_i = 0$, and therefore $y_r' = 0$ $(r = 1, 2, 3)$. One infers that when there is a cavity, the maximum of $|Y_r|$ is larger than the value $|p_e|$ assumed by it for a solid body.

2. The work done by the external forces is equal to $- p_e \Delta V_e + p_i \Delta V_i$. Then, from (6.77), (6.103), (6.104) it follows that

$$- p_e \Delta V_e + p_i \Delta V_i > \frac{1}{C^*} [A (p_e V_e - p_i V_i)^2 + B V_e^2 (p_e - p_i)^2], \quad (6.107)$$

where

$$A = \sum_{r,s=1}^{3} m_{rs} > 0, \qquad B = \sum_{t=1}^{3} \frac{1}{\varrho_t^2} \sum_{r,s=1}^{3} m_{rs} y_r' y_s' > 0. \quad (6.108)$$

From the expression of Hooke's Law,

$$\varepsilon_r = - \sum_{s=1}^{6} m_{rs} Y_s, \quad (6.109)$$

one deduces that the variation of the volume of the shell is expressed by

$$\Delta V_e - \Delta V_i = - \sum_{r=1}^{3} \sum_{s=1}^{6} m_{rs} \int_{C^*} Y_s dC^* = A (p_i V_i - p_e V_e). \quad (6.110)$$

From (6.107), (6.110) one derives the inequalities

$$\Delta V_e \lessgtr \frac{V_e}{C^*} [A (p_i V_i - p_e V_e) + B V_e (p_i - p_e)],$$

$$\Delta V_i \lessgtr \frac{1}{C^*} [A V_i (p_i V_i - p_e V_e) + B V_e^2 (p_i - p_e)], \quad \left.\right\} \quad (6.111)$$

where the sign $>$ or $<$ is to be taken according as $p_i - p_e$ is positive or negative. From (6.111) it follows that:

a) If $p_i - p_e < 0$, the right-hand members of (6.111) are negative, and the volumes V_i, V_e decrease by quantities which are greater than the second members of (6.111), whatever be the constitution of the homogeneous body, provided the external forces applied on σ_i be pressures $[p_i \geqslant 0]$, or, if $p_i < 0$, provided the difference $q = p_i - p_e$ satisfy the inequality

$$q < p_i \frac{V_e - V_i}{V_e}. \tag{6.112}$$

b) If $p_i - p_e > 0$, the second members of (6.111) are positive and V_i, V_e increase, whatever be the constitution of the body, provided the set of external forces acting on σ_e be tractions $[p_e \leqslant 0]$ or, if $p_e > 0$, provided that

$$q > p_e \frac{V_e - V_i}{V_i}. \tag{6.113}$$

In the remaining possibilities beyond those included in cases a), b), it being supposed that $B > 0$, one may have dilatation or contraction of the volumes V_e, V_i, according to the constitution of the continuous medium. For example, if $p_i - p_e > 0$, $p_e > 0$, the second members of (6.111) are positive even if (6.113) is not satisfied but instead

$$q > \frac{A(V_e - V_i)}{AV_i + BV_e} p_e. \tag{6.114}$$

In this case V_e and V_i increase by more than the values of the second members of (6.111).

Further, if $p_i - p_e > 0$, $p_e > 0$, the second member of (6.111, 2) is positive when B is positive, even if (6.114) is not satisfied but instead

$$q > \frac{AV_i(V_e - V_i)}{AV_i^2 + BV_e^2} p_e. \tag{6.115}$$

In this case (6.111) show that V_i increases, and the right-hand member of (6.111, 2) gives a lower bound for its increase.

3. If C^* has several cavities of volumes V_1, V_2, $\cdots$, V_n, bounded by surfaces σ_1, σ_2, $\cdots$, σ_n, while σ_e is the external boundary and V_e the volume enclosed, one finds that

$$\sum_{r,s=1}^{6} \int_{C^*} q_{rs} Y_r Y_s dC^* \geqslant \frac{1}{C^*} \sum_{r,s=1}^{3} q_{rs} \left\{ \left[p_e V_e - \sum_{\nu=1}^{n} p_\nu V_\nu \right]^2 + \right.$$

$$\left. + \sum_{\nu=1}^{n} (p_e - p_\nu)^2 V_\nu^2 \sum_{t=1}^{3} \frac{y_r^{(\nu)} y_s^{(\nu)}}{\varrho_t^2} \right\}, \tag{6.116}$$

where $y_r^{(\nu)}$ ($r = 1, 2, 3; \nu = 1, 2, \cdots, n$) are the coordinates of the center of mass of V_ν.

It is easy to see that if the pressures acting on the boundaries of V_1, $V_2, \cdots V_n, V_e$ are all equal except one, from (6.116) it is possible to deduce an inequality regarding the variation of the volume within that boundary.

§ 12. New inequalities regarding the displacements

1. The previous inequalities for the displacements are profitable only if it is possible to make evident, in the expression of work done by the external forces, the characteristic elements of the deformation. But by using Hooke's Law it is possible to establish other inequalities for which that is not necessary. In fact, from

$$Y_{rs} = - [\lambda \, \delta_{rs} \, u_p^{;p} + \mu \, (u_{r,s} + u_{s,r})] \tag{6.117}$$

it follows that

$$\mu \, (u_{r,s} + u_{s,r}) = \frac{\lambda}{3\lambda + 2\mu} \, \delta_{rs} \, Y_r^{\cdot r} - Y_{rs} \tag{6.118}$$

and

$$\mu \int_{C^*} (u_{r,s} + u_{s,r}) \, Q_t \, dC^* = - \frac{C^*}{3\lambda + 2\mu} \, [(3\lambda + 2\mu) \, \overline{Y_{rs} Q_t} - \lambda \delta_{rs} \, \overline{Y_r^{\cdot r} Q_t}] , \tag{6.119}$$

where Q_t is an arbitrary function of y_1, y_2, y_3.

Let us suppose that the displacements are assigned on the portion Σ_1^* of Σ^*, while the surface forces are known on the remainder $\Sigma_2^* = \Sigma^* - \Sigma_1^*$.

From (6.119) it follows that

$$\mu \{ |u_r|_{\max} [\int_{C^*} |Q_{t,s}| \, dC^* + \int_{\Sigma_1^*} |Q_t N_s^*| \, d\Sigma_2^*] + |u_s|_{\max} [\int_{C^*} |Q_{t,r}| \, dC^* +$$

$$+ \int_{\Sigma_1^*} |Q_t N_r^*| \, d\Sigma_2^*] \} \geqslant \left| C^* [\lambda \delta_{rs} \, \overline{Y_p^{;p} Q_t} - (3\lambda + 2\mu) \, \overline{Y_{rs} Q_t}] \frac{1}{3\lambda + 2\mu} - \right.$$

$$\left. - \mu \int_{\Sigma_1^*} (u_r N_s^* + u_s N_r^*) \, Q_t \, d\Sigma_1^* \right| . \tag{6.120}$$

(6.120) may be useful to determine lower bounds for $|u_r|_{\max}$. In particular, for $r = s$ one has

$$2\mu \, |u_r|_{\max} \geqslant$$

$$\frac{|C^* [\lambda \, \overline{Y_p^{;p} Q_t} - (3\lambda + 2\mu) \, \overline{Y_{rr} Q_t}] - 2\mu (3\lambda + 2\mu) \int_{\Sigma_1^*} u_r Q_t N_r^* \, d\Sigma_1^*|}{(3\lambda + 2\mu) [\int_{C^*} |Q_{t,r}| \, dc^* + \int_{\Sigma_1^*} |Q_t N_r^*| \, d\Sigma_2^*]} . \tag{6.121}$$

Evidently, if Q_t does not depend upon y_r, (6.121) gives a lower bound on Σ_2^* for $|u_r|_{\max}$. If the second member of (6.120) is known, profit may be drawn from the corresponding inequality. That happens if Q_t is a linear combination with constant coefficients α_i,

$$Q_t = \sum_{i=0}^{t} \alpha_i P_i , \tag{6.122}$$

of polynomials chosen so that $\overline{Y_{rs}Q_t}$ are all or partially known, according to what was explained in Sections 2 and 3. If the $\overline{Y_{rs}Q_t}$ are not all known, one may render known the second member of (6.120) by giving to un-determined elements values such as to make it a minimum.

But, if the external forces are all known, Σ_1^* vanishes, the integral cor-responding to Σ_1^* is zero in (6.120), and the polynomials P_i may be chosen so that all $\overline{Y_{rs}P_t}$ are known. In particular, that happens if one sets $t = 3$ and $P_0 \equiv 1$, $P_i = y_i$. In this case $\overline{Y_{rs}Q_t}$ is a linear combination of astatic and bi-astatic coordinates.

2. I shall show a very simple application of (6.121) by considering a problem analogous to the one in Section 10. Let C^* be a body bounded by two plane bases σ^+, σ^-, equal and parallel, and by a lateral surface σ_l. I suppose that σ^+, σ^- are perpendicular to the axis of y_3 and that the exter-nal forces f^+ (y_1, y_2) and f^- (y_1, y_2) are parallel to the axis of y_3 and act only on σ^+, σ^- respectively. Then

$$\overline{Y_{33}} = -\frac{2R}{C^*} l , \qquad Y_{rs} = 0 \qquad (r \text{ or } s \neq 3) , \tag{6.123}$$

where R denotes the component along the y_3 axis of the resultant of f^*, and $2\,l$ denotes the distance between σ^+ and σ^-. Plainly

$$\int_{\Sigma^*} |N_3^*| \, d\Sigma^* = 2 A' \tag{6.124}$$

where A' denotes the area inclosed by the projection of the boundary of the body upon the plane $y_3 = 0$.

For $Q_t \equiv 1$, from (6.121) it follows that

$$|u_3|_{\max} > \frac{(\lambda + \mu)\, l\, |R|}{(3\lambda + 2\mu)\, \mu\, A'} . \tag{6.125}$$

Instead, for the transversal contraction it may be shown from (6.121) that

$$|u_i|_{\max} > \frac{l\, |\lambda R|}{2\mu\, (3\lambda + 2\mu)\, A_i'} \qquad (i = 1, 2), \tag{6.126}$$

where A_i' denotes the area inclosed by the projection of the boundary of the body upon the plane $y_i = 0$.

Not much can be claimed for (6.125), (6.126), which are obtained from (6.121) in such a simple way, only by making $Q_t \equiv 1$, but (6.125), (6.126) may be considered as generalizations of the well known formula for simple extension according to SAINT-VENANT's theory. If one denotes Poisson's coefficient by v, from (6.125), (6.126) it follows that quotient φ_i of the second members of those equations is

$$\varphi_i = |v| \frac{A'}{A'_i} \qquad (i = 1, 2), \tag{6.127}$$

Then, if the second members of (6.125), (6.126) are assumed as approximate values of $|u_3|_{max}$, $|u_i|_{max}$, the quotient of the maximum transversal contraction and the maximum longitudinal displacement is expressed by φ_i. This value becomes equal to $\varphi_i = \pi \dfrac{|v| \, r}{4 \, l}$ in the case of a cylindrical body of radius r and is slightly less than the one given by SAINT-VENANT's theory.

§ 13. On the variation of temperature in an adiabatic transformation

From the general inequalities (6.16), (6.22) it is possible to deduce lower bounds for the variation of temperature corresponding to a state of adiabatic equilibrium. To this end, it is sufficient to denote by τ the linear form in Y_r by which the variation of temperature is expressed and to take q_{rs} in (6.16) and β_s in (6.22) in such a manner that

$$\tau = \sqrt{\sum_{r,s=1}^{6} q_{rs} Y_r Y_s} = \sum_{s=1}^{6} \beta_s Y_s . \tag{6.128}$$

Supposing the body isotropic, let λ and μ be Lamé's coefficients, L the coefficient of thermal tension and C another coefficient which is proportional to the quotient of the specific heat at constant configuration by the absolute temperature in C^*. Put

$$I_1(Y) = \sum_{i=1}^{3} Y_i. \tag{6.129}$$

From (6.16) it follows that

$$|\tau|_{max} \geqslant \frac{L}{\left[3\left(\lambda + \dfrac{L^2}{c}\right) + 2\mu \right] c} \sqrt{\sum_{i=0}^{m} \frac{1}{\varrho_i^2} \overline{I_1(Y) P_i}^2} \geqslant$$

$$\geqslant \frac{L}{\left[3\left(\lambda + \dfrac{L^2}{c}\right) + 2\mu \right] c} \sqrt{\overline{I_1^2(Y)} + \sum_{i=1}^{3} \frac{\overline{I_1(Y) y_i}^2}{\varrho_i^2}}, \tag{6.130}$$

while from (6.22) one deduces that

$$
|\tau|_{max} \geq \frac{C*L \sum_{t=0}^{m} \frac{1}{\varrho_t^2} \overline{I_1(Y)\,P_t}^{\,2}}{\left[3\left(\lambda + \frac{L^2}{c}\right) + 2\mu\right]c \int_{C*} \left| \sum_{t=0}^{m} \frac{1}{\varrho_t^2} \overline{I_1(Y)\,P_t} \cdot P_t \right| dC*} \geq
$$

$$
\geq \frac{C*L\left[I_1^2(Y) + \sum_{t=1}^{3} \frac{\overline{I_1(y)\,y_t}^{\,2}}{\varrho_t^2}\right]}{\left[3\left(\lambda + \frac{L^2}{c}\right) + 2\mu\right]c \int_{C*} \left| I_1(Y) + \sum_{t=1}^{3} \frac{1}{\varrho_t^2} \overline{I_1(Y)\,y_t}\,y_t \right| dC*} . (6.131)
$$

In (6.130), (6.131) the third members are known functions of the astatic and hyper-astatic coordinates of the external forces.

§ 14. Upper bounds for displacements and stresses

The inequalities obtained in the previous sections give lower bounds which are useful in establishing necessary conditions of safety. Instead, upper bounds are necessary for sufficient conditions of safety. The problem of determining such inequalities, generally, is clearly more difficult and may have no solution, as happens, for example, if there is a stress singularity. In this section I only indicate some upper bounds[1] [G. SUPINO], distinguishing two cases.

Case a). First boundary-value problem. Let us suppose the displacements are assigned on the whole boundary $\Sigma*$ and the body forces all zero. Let $G(P*,Q*)$ be the Green's function, $P*$ being the typical point in $C*$ and $Q*$ the typical point on $\Sigma*$. Put

$$
\Delta = u_{,r}^{\,r} . \tag{6.132}
$$

From the equations of equilibrium, written in the form proposed by TEDONE, it follows that

$$
u_r = -\frac{\lambda + \mu}{2\mu}\,y_r\,\Delta + \frac{1}{4\mu} \int_{\Sigma*} \frac{\partial G(P*,Q*)}{\partial N_{Q*}^*}\,u_r(Q*)\,d\Sigma* +
$$

$$
+ \frac{\lambda + \mu}{8\pi\mu} \int_{\Sigma*} \frac{\partial G(P*,Q*)}{\partial N_{Q*}^*}\,y_r(Q*)\,\Delta(Q*)\,d\Sigma* . \tag{6.133}
$$

---

[1] For other upper and lower bounds for the displacements in the first boundary-value problem, the surface displacement being given, see DIAZ and GREENBERG. Other upper and lower bounds have been obtained by SYNGE for the displacement and the dilatation in the first boundary-value problem and for the stress components and the sum of the principal stresses in the second boundary-value problem, the surface stress being given.

Plainly, if ones gives an upper bound for $|\Delta(Q^*)|$ on Σ^*, it is possible to derive from (6.133) an upper bound for $|u_r(P^*)|$. Let

$$\eta_i(P^*) = \frac{1}{4\pi} \int_{\Sigma^*} \frac{\partial G(P^*,Q^*)}{\partial N^*_{Q^*}} u_i(Q^*) d\Sigma^* . \tag{6.134}$$

Supposing

$$|\eta^i_{,i}| \leqslant K \qquad (K > 0) , \tag{6.135}$$

it is possible to demonstrate that

$$|\Delta(P^*)| \leqslant K . \tag{6.136}$$

If $\eta^i_{,i}$ is negligible except on a small portion Σ' of Σ^*, a better inequality may be substitued for (6.136). For example, if $\Sigma' = \frac{1}{20} \Sigma^*$ and Poisson's coefficient $\nu = \frac{1}{4}$, one has

$$\Delta(P^*) \begin{cases} \leqslant \dfrac{K}{2}\left(1 + \dfrac{\pi^2}{40}\right) & \text{(on } \Sigma'\text{)} , \\[2mm] < \dfrac{K}{8} & \text{(on } \Sigma^* - \Sigma'\text{)} . \end{cases} \tag{6.137}$$

A sharper upper bound may be obtained for $|u_r(P^*)|$ by suitably increasing $\left|\dfrac{\partial G(P^*,Q^*)}{\partial N^*_{Q^*}}\right|$.

Case b). Plane problem with assigned forces on the boundary. Let us consider a plane medium referred to coordinate axes $O^* y_1 y_2$ and subjected to external forces with components f_1, f_2, applied to the boundary Σ^*. The only non-vanishing stress components are Y_{11}, Y_{22}, Y_{12}. Let us suppose that $|f_1^*|$ and $|f_2^*|$ are smaller than f^* on a portion Σ' of Σ^* while on $\Sigma^* - \Sigma'$ they are zero. By using Airy's function and by considerations partially analogous to those of Case a) it is possible to demonstrate [Supino] that when Σ'_1 is rather small, at any point P^* rather far from Σ' the following inequality holds:

$$Y_{11}(P^*) < \frac{\pi f^* \Sigma'^2}{8r^2} + \frac{f^* \Sigma' l}{2\sqrt{2}\,r}\left[\cos(Q^*P^*, N^*_{Q^*}) + \right.$$

$$\left. + \frac{\pi^2}{8}\cos(y_1, r)\right] + \frac{\bar{K}\pi}{8}\int_{\Sigma^*}\left|\frac{\cos(y_1, r)}{r}\right| d\Sigma^* + \frac{\bar{K}}{2} , \tag{6.138}$$

where l^* denotes the length of the boundary and $r = |Q^*P^*|$. $\bar{K}$ is a constant which is to be found. If the boundary is a circle of radius R,

(6.138) becomes

$$Y_{11}(P^*) < \frac{\pi f^* \Sigma'^2}{8} + \pi f^* \Sigma' R \left[\frac{\sqrt{2}}{r} \cos (Q^* P^*, N^*_{Q^*}) + \right.$$

$$+ \frac{3}{R} \left(\sqrt{2} + \frac{1}{4} \right) + \frac{\pi^2}{4\sqrt{2}\,r} \cos (y_1, r) \bigg] +$$

$$+ 3\,\pi f^* \Sigma' \left[\frac{\pi}{32} \int\limits_{\Sigma^*} \left| \frac{\cos(y_1, r)}{r^2} \right| d\Sigma^* + \frac{1}{\sqrt{2}} \right], \qquad (6.139)$$

Inequalities analogous to (6.138), (6.139) hold for the other non-zero stresses.

§ 15. Other upper bounds for the stresses

1. Interesting upper bounds have been established by G. COLOMBO [2] for the equilibrium of cylindrical elastic bodies according to SAINT-VENANT's theory. Specifically, they concern[1] the problems of torsion and of non-uniform bending. I shall report on them briefly.

a) *Torsion problem.* Let us suppose that the $O^* y_3$ axis of the central reference frame is parallel to the generators of the cylinder. Let A^* be the cross section, and let σ_l be the boundary of A^*. As is well known, according to SAINT-VENANT's theory the torsion problem is formulated by the equations

$$Y_{11} = Y_{22} = Y_{33} = Y_{12} = 0 , \qquad (6.140)$$

$$Y_{13} = \frac{\partial \Phi}{\partial y_2} , \quad Y_{23} = -\frac{\partial \Phi}{\partial y_1} , \qquad (6.141)$$

$$\left. \begin{array}{ll} \Delta_2 \Phi = -K & \text{(in } A^*) , \\[2mm] \Phi = 0 & \text{(on } \sigma_l) , \end{array} \right\} \qquad (6.142)$$

$$2 \int\limits_{A^*} \Phi \, dy_1 \, dy_2 = M_t , \qquad (6.143)$$

where M_t is the twisting moment of the external forces applied to one of the bases. Supposing A^* to be a simply connected region, let E_1, $E_2, \cdots, E_n$ be n ellipses without internal points in common and with no point exterior to A^*. It may be supposed that $M_t \geqslant 0$, $K \geqslant 0$. Then, necessarily, $\Phi \geqslant 0$ in A^*, and

$$\int\limits_{A^*} \Phi \, dy_1 \, dy_2 \geqslant \sum_{r=1}^{n} \int\limits_{E_r} \Phi \, dy_1 \, dy_2 . \qquad (6.144)$$

---

[1] Other upper bounds have been established by G. COLOMBO [1] for the problem of plane deformation.

In E_r one may suppose that

$$\Phi = u_r + v_r \tag{6.145}$$

where u_r, v_r are two functions which satisfy equations

$$\left.\begin{array}{llll}
\Delta_2 u_r = -K & \text{(in } E_r\text{)} & u_r = 0 & \text{(on } \sigma_r\text{)} \\
\Delta_2 v_r = 0 & \text{(in } E_r\text{)} & v_r = \Phi & \text{(on } \sigma_r\text{)}
\end{array}\right\} \tag{6.146}$$

and σ_r is the boundary of E_r. Denoting by a_r, b_r the semi-axes of E_r, we obtain

$$u_r = -\frac{K a_r^2 b_r^2}{2 (a_r^2 + b_r^2)}\left(\frac{x_r^2}{a_r^2} + \frac{\xi_r^2}{b_r^2} - 1\right), \tag{6.147}$$

where x_r, ξ_r are the coordinates of the typical point of E_r with respect to its axes of symmetry. Further, since $\Phi_r \geqslant 0$, from (6.146, 2) it follows that $v_r \geqslant 0$. From (6.143), (6.144), (6.145), (6.147) and from $v_r \geqslant 0$ within E_r, it follows that

$$0 < K \leqslant \frac{2 M_t}{\pi}\left[\sum_{r=1}^{n} \frac{a_r^3 b_r^3}{a_r^2 + b_r^2}\right]^{-1}, \tag{6.148}$$

which gives a lower bound for the torsional rigidity $\dfrac{M_t}{K}$. Let us suppose that the boundary σ_l of $A*$ is given by the equation

$$\varrho = \alpha(\theta), \tag{6.149}$$

where $\alpha(\theta)$ is a positive periodic function with period 2π. Further, let us suppose that $\alpha(\theta)$ has first, second and third derivatives and that the radius of curvature of σ_l is never zero. Let us denote by ϱ_0 the radius of the greatest circle contained in $A*$ and by R_0 the maximum value of the radius of curvature of σ_l. If ϱ_1, ϱ_2 are the minimum and maximum values of $\alpha(\theta)$, ϱ' the maximum of $\left|\dfrac{d\alpha}{d\theta}\right|$ and ϱ_1'', ϱ_2'' the minimum and maximum of $\left|\dfrac{d^2\alpha}{d\theta^2}\right|$, one may demonstrate that

$$|Y_{13}| \leqslant \tau, \qquad |Y_{23}| \leqslant \tau, \tag{6.150}$$

where

$$\tau = \frac{K'}{4}\left[\frac{\varrho_2^2 + \varrho_1\varrho_2 - \varrho_1^2}{\varrho_1} + \frac{\varrho'^2}{\varrho_1}\left(1 + \frac{\varrho_2^2}{\varrho_1^2}\right) + \frac{\varrho_0}{2}\left(2 - \frac{\varrho_0}{R_0}\right)\frac{\varrho_2''}{\varrho_1}\left(1 + \frac{\varrho_2^2}{\varrho_1^2}\right)\right] \tag{6.151}$$

and where K' is expressed by the third member of (6.148). It is easily verified that the sign of equality is valid in (6.150) if the section $A*$ is a circle.

6. *Non-uniform bending.* Now, let us suppose that the set of external forces acting upon one of the cylinder's bases is equipollent to one force f^* applied at a point on that base and having the same direction as the y_1-axis. Let the axes y_1, y_2 coincide with the principal axes of the base. Denoting Poisson's coefficient by ν and the moment of inertia of A^* with respect to the y_2-axis by I, let us put

$$\lambda' = \frac{3 + 2\nu}{8(1 + v)}, \qquad r^2 = \frac{\varrho_1^2 + \varrho_2^2}{2}, \tag{6.152}$$

$$\begin{aligned}
\overline{\Psi}_0 &= \frac{f^*}{I} \left[\frac{\lambda'}{2} \varrho_2 (\varrho_2^2 - \varrho_1^2) + 4 \varrho_2^2 \varrho' \right], \\[2mm]
\overline{\Psi}_0' &= \frac{f^*}{I} \left\{ \left[\frac{1 + 6\nu}{8(1 + \nu)} \varrho_1^2 + \lambda' \varrho_2^2 \right]^2 + \frac{\lambda'^2}{4} (\varrho_2^2 - \varrho_1^2)^2 \right\}^{\frac{1}{2}}, \\[2mm]
\overline{\Psi}_0'' &= \frac{f^*}{I} \left\{ 2(3\lambda' - 1) \varrho_2 \varrho'^2 + (2\lambda' - 1) \varrho_2 \varrho_2'' + \right. \\[1mm]
&\quad \left. + \frac{\lambda}{2} (\varrho_2^2 - \varrho_1^2)(\varrho'' + \varrho_2) + \left[\frac{10\lambda' - 1}{2} \varrho_2^2 - \lambda' \varrho_1^2 \right]^2 \varrho_1'^2 \right\}^{\frac{1}{2}}, \\[2mm]
\overline{\Psi}_n &= \frac{1}{\varrho_1} \left(\varrho' \frac{\Psi_0'}{\varrho_1} + 2\overline{\Psi}_0 \right) + \frac{\varrho_0 \varrho_2}{2 \varrho_1^2} \left(2 - \frac{\varrho_0}{R_0} \right) (\Psi_0'' + 2\varrho_2'' \overline{\Psi}_0 + \\[1mm]
&\quad + 2\varrho'^2 \Psi_0'), \\[2mm]
\overline{\Psi}_p &= \left[\frac{\Psi_0'}{\varrho_1^2} + \overline{\Psi}_n^2 \right]^{\frac{1}{2}}.
\end{aligned} \right\} \tag{6.153}$$

If the straight line along which f^* is applied contains the center of torsion of A^*, one may demonstrate the inequalities

$$\left.\begin{aligned}
|Y_{13}| &\leqslant \frac{3 + 2\nu}{8(1 + \nu)} \frac{f^*}{I} \left| r^2 - y_1^2 - \frac{1 - 2\nu}{3 + 2\nu} y_2^2 \right| + \overline{\Psi}_p, \\[2mm]
|Y_{23}| &\leqslant \frac{1 + 2\nu}{4(1 + \nu)} \frac{f^*}{I} |y_1 y_2| + \overline{\Psi}_p,
\end{aligned}\right\} \tag{6.154}$$

while, as is well known,

$$Y_{33} = \frac{f^*}{I} (y_3 - l) y_1, \qquad Y_{11} = Y_{22} = Y_{12} = 0. \tag{6.155}$$

In the case of a circular section $\overline{\Psi}_p = 0$, and the sign of equality is valid in (6.154).

If the straight line along which f^* is applied does not contain the center of torsion C_t of A^*, and if d is the distance from C_t to the center of mass of A^*, we obtain the inequality

$$0 \leqslant d < \frac{1}{I} \left\{ \frac{1}{4(1 + \nu)} \int_{A^*} [2 y_1^2 y_2 - (1 - 2\nu) y_2^3] \, dy_1 dy_2 + \right.$$

$$\left. + 3 A^* \left[\frac{\lambda' \varrho_2}{2} (\varrho_2^2 - \varrho_1^2) + 4 \varrho_2^2 \varrho' \right] \right\}. \tag{6.156}$$

The inequalities (6.154) are valid when $d = 0$. If $d > 0$, the inequalities corresponding to (6.154) are obtained by adding to the second members of (6.154) the second members of (6.150), after having here identified M_t with the sum of $f*\delta$ and the product of $f*$ by the third member of (6.156), δ being the distance between the straight line along which $f*$ is applied and the center of mass of $A*$.

Chapter VII

Integration of the Fundamental Problem of Static Elasticity

§ 1. Preliminaries

1. In the previous chapter I have established integral properties of the equations of equilibrium of a continuous medium with respect to the sequence of monomials constructed with the Cartesian coordinates, showing that it is possible to derive some inequalities based on such properties. But their scope is more general, for it is possible to set up an integration method based on those integral properties and valid for homogeneous, isotropic or anisotropic, slightly deformable bodies (GRIOLI [5]). By this method the principal unknowns are just the six components of stress Y_{rs}, but the fact that these, rather than the displacements, are the principal unknowns does not seem to be disadvantageous: almost always knowledge of the stress and of the forces prescribed by the constraints is more interesting than knowledge of the displacements, while it is well known that any method giving the displacements directly in the form of a power series generally has the disadvantage that the convergence of the derivatives of the power series—by which stress and constraint forces are expressed—is numerically slow, and the number of terms that one may have calculated is not sufficient to approximate well the derivatives of the power series. Instead, this inconvenience is absent if one determines displacement from the stresses by an integration procedure.

The fact that the polynomials by which stress components are approximated, according to the method I will set out, satisfy SAINT-VENANT's integrability conditions, whatever the number of terms considered, is very useful, as it is then possible—and easy—to determine the corresponding approximations to the displacement components.

The method is applicable, with some care, also in the case of *unilateral*[1] constraints and in that of elastic dislocations, but it presupposes a complete characterization of the kinds of forces which may be prescribed by constraints. For simplicity, I shall consider only the cases of clamping or supporting constraints.

---

[1] See footnotes (1) on pp. 40, 60.

The expressions obtained for stress and for the forces of constraint satisfy in mean the equations of equilibrium corresponding to any portion of the region occupied by the body and the boundary. But, analyzing the procedure by which those equilibrium equations are deduced, one recognizes that no more than this is required unless special regularity conditions for the unknowns are presumed, conditions which cannot be deduced from the basic equations of mechanics. What I shall say refers to a homogeneous isotropic or anisotropic, slightly deformable elastic body whose elastic potential energy density is expressed by a positive-definite quadratic form in the stress components. Further, it is supposed that the solution exists and is unique and the Y_{rs} are square-integrable in the region occupied by the body.

2. To make what follows clear it is well to note that the analytic problem to be studied consists in the choice of that solution of the equations

$$Y_{rs}^{;s} = k^* F_r \qquad\qquad \text{(in } C^*\text{)} , \qquad\qquad (7.1)$$

$$Y_{rs} N_*^s = f_r^* \qquad\qquad \text{(on } \Sigma^*\text{)} \qquad\qquad (7.2)$$

which is integrable for the determination of the displacement components. The surface forces f^* are unknown on the portion Σ_1^* where constraints may be present. The stated aim may be reached by the application of Menabrea's theorem. Then the problem consists in choosing, among all solutions of (7.1), (7.2), the one which minimizes the expression

$$V(Y) - \int_{\Sigma_1^*} \Phi_r u^r d\Sigma_1^* , \qquad\qquad (7.3)$$

where

$$V(Y) = \frac{1}{2} \int_{C^*} \sum_{r,s=1}^{6} m_{rs} Y_r Y_s dC^* \qquad\qquad (7.4)$$

is the elastic potential energy expressed by means of Y_r and where Φ_r is the r-component of the force corresponding to the constraints at a typical point of Σ_1^* and u_r is the analogous displacement component at the same point.

The minimum of expression (7.3) is to be found when the Y_{rs} vary in the class of all stress states which satisfy (7.1) and (7.2) on $\Sigma_2^* = \Sigma^* - \Sigma_1^*$ and when the Φ_r vary in the class of every possible constraint force in equilibrium with the given external forces. After having determined the solution which makes expression (7.3) a minimum, the displacement components are determined by integration of well known expressions constructed by means of ε_{rs}, namely,

$$\varepsilon_r = -\frac{\partial W(Y)}{\partial Y_r} = -\sum_{s=1}^{6} m_{rs} Y_s \qquad (r = 1, 2 \cdots, 6) . \qquad (7.5)$$

Note that it is to be presumed that the analytic problem has no solution if the u_r, determined in such manner, do not satisfy the constraint conditions, keeping in mind arbitrary elements contained in the u_r. Otherwise, the solution determined by MENABREA's theorem *must* verify the constraint conditions, *if a solution exists.*

However, it is well to note that the minimum of (7.3) is to be determined *after having characterized in precise and complete manner the set of the possible forces that the constraints are capable of prescribing.*

For example, if the body is supported on a portion Σ_1', of Σ_1^* without friction, the constraint forces, if not zero, are perpendicular to the planes tangent to Σ_1'. This fact is expressed by the equations

$$\left.\begin{array}{c} \Phi_r = \varrho N_r^* \\[2mm] \varrho \geqslant 0 \end{array}\right\} \text{ (on } \Sigma_1') . \qquad (7.6)$$

It is well to observe that if the body is clamped or supported without friction on Σ_1^*, or if constraints are absent, the expression which must be minimized is not (7.3) but (7.4), since either $u_r = 0$ on Σ_1^*, or the displacement is perpendicular to Φ if this is different from zero, or Σ_1^* vanishes. Naturally, in the case of a unilateral support it is necessary to keep in mind the possibility of detachment on a portion of the surface of support, but I shall take up this matter in § 4.

§ 2. Analytical preface to the integration of the static elasticity problem

Let $\{w_t\}$ be the sequence of all possible monomials formed by means of y_1, y_2, y_3 and placed in non-decreasing degree as t increases, and let P_t be the polynomial[1] obtained by adding w_t to that linear combination of $w_0, w_1, \cdots w_{t-1}$ which makes P_t orthogonal to $w_0, w_1, \cdots w_{t-1}$ in C^*. Let us denote by $\{P_t\}$ the sequence of all linearly independent polynomials so defined, orthogonal in C^*. Further, let u and Ψ be two functions of y_1, y_2, y_3 defined in C^*, φ a function defined on Σ^*, and let

$$E(u) = a^{ih} u_{,ih} + b^i u_{,i} + cu = \Psi \qquad (7.7)$$

be a second-order elliptic equation with constant coefficients. Let $E^*(u)$ be the differential expression adjoint to $E(u)$.

Specializing some results of AMERIO, one has: *the functions u, Ψ, φ satisfy* (7.7) *almost everywhere and satisfy also the conditions*

$$u = 0, \qquad \frac{du}{dN^*} = \varphi \qquad \text{(on } \Sigma^*) , \qquad (7.8)$$

---

[1] Such a procedure of orthogonalization is certainly possible by Gram's theorem (PICONE [3]).

if for each term of the sequence $\{w_t\}$ the equation

$$\int_{C^*} u\, E^*(w_t)\, dC^* = \int_{C^*} \Psi w_t\, dC^* + \int_{\Sigma^*} \varphi w_t\, d\Sigma^* \qquad (t = 0, 1, \cdots,) , \qquad (7.9)$$

is satisfied.

From AMERIO's theorem it follows that *the functions Ψ and φ are equal to zero in C^* and on Σ^*, respectively, if they satisfy the Fischer-Riesz equations*

$$\int_{C^*} \Psi w_t\, dC^* + \int_{\Sigma^*} \varphi w_t\, d\Sigma^* = 0 \qquad (t = 0, 1 \cdots) . \qquad (7.10)$$

To see this it is sufficient to observe that under hypothesis (7.10), (7.9) is satisfied for $u \equiv 0$ and to bear in mind (7.7), (7.8). By the above *completeness* property of the sequence $\{w_t\}$ one deduces, clearly, the completeness of any sequence of all linearly independent polynomials which may be represented by linear combinations of w_t.

In particular, this happens for the sequence $\{P_t\}$. Therefore one has: *the equations*

$$\int_{C^*} \Psi P_t\, dC^* + \int_{\Sigma^*} \varphi P_t\, d\Sigma^* = 0 \qquad (t = 0, 1 \cdots) , \qquad (7.11)$$

are necessary and sufficient conditions that Ψ and φ vanish almost everywhere in C^* and on Σ^*, respectively.

From this completeness property of $\{P_t\}$ follows the well known property that any square-integrable function in C^* is expressible by a power series in P_t which converges in the mean.

From the completeness property of the succession $\{P_t\}$ follow:

a) *If two sets of external forces characterized respectively by vectors F, f^* and F', f' satisfy the infinite set of equations*

$$\int_{C^*} F P_t\, dC^* + \int_{\Sigma^*} f^* P_t\, d\Sigma^* = \int_{C^*} F' P_t\, dC^* + \int_{\Sigma^*} f' P_t\, d\Sigma^* \qquad (t = 0, 1 \cdots) ,$$
$$(7.12)$$

they coincide almost everywhere.

b) *If equations (7.1) admit square-integrable solutions, these are expressible by the series*

$$Y_r = \sum_{t=0}^{\infty} \frac{\overline{Y_r P_t}}{\varrho_t^2} P_t \qquad (r = 1, 2, \cdots, 6) . \qquad (7.13)$$

§ 3. Polynomial approximations of stress components

The external forces are generally unknown on a portion Σ_1^* of the boundary Σ^* of C^* if there are constraints. Put

$$c_{\eta \tau \lambda}^{(r)} = -\frac{1}{C^*} \int_{\Sigma_1^*} y_1^{\eta} y_2^{\tau} y_3^{\lambda} \Phi^r\, d\Sigma_1^* . \qquad (7.14)$$

The $b_{\eta\tau\lambda}^{(r)}$ [see (6.3)] are the sum of a known term corresponding to the given forces and of one generally unknown, $c_{\eta\tau\lambda}^{(r)}$. Naturally, Σ_1^* vanishes and the $c_{\eta\tau\lambda}^{(r)}$ are zero if constraints are absent.

Let us denote by G_m a possible choice of $m+1$ sextuples $\overline{Y_r P_t}$ $(t = 0, 1, \cdots, m)$ of class M [see Chap. VI, § 2, at the end] corresponding to a prescribed choice of $c_{\eta\tau\lambda}^{(r)}$, and by I_m the set of all possible choices G_m considered also when the $c_{\eta\tau\lambda}^{(r)}$ vary in such a way as to satisfy the basic equations of statics and be compatible with the nature of the constraints.

For example, if the body is supported without friction on a portion Σ_1^* of Σ^*, the vector $\boldsymbol{\Phi}$ by which the $c_{\eta\tau\lambda}^{(r)}$ are constructed, according to (7.14), is parallel to $\boldsymbol{N^*}$. This condition is easily expressible analytically if Σ_1^* is a portion of a plane: it is sufficient to suppose one of the reference frame axes, for example y_3, perpendicular to Σ_1^* and oriented towards the region of space which contains the body, since from this follows $\boldsymbol{\Phi}_1 = 0$, $\boldsymbol{\Phi}_2 = 0$, $\boldsymbol{\Phi}_3 > 0$, and then all $c_{\eta\tau\lambda}^{(r)}$ are equal to zero except $c_{\eta\tau0}^{(3)}$ [see also Observation I, § 4].

Bearing in mind that

$$W(Y) = \frac{1}{2} \sum_{r,s=1}^{6} m_{rs} Y_r Y_s \tag{7.15}$$

is the density of isothermal potential elastic energy, let us put

$$Y_r^{(m)} = \sum_{t=0}^{m} \frac{\overline{Y_r^{(m)} P_t}}{\varrho_t^2} P_t , \tag{7.16}$$

$$V_m = \frac{1}{2} \sum_{t=0}^{m} \frac{1}{\varrho_t^2} \sum_{r,s=1}^{6} m_{rs} \overline{Y_r^{(m)} P_t} \; \overline{Y_s^{(m)} P_t} . \tag{7.17}$$

It is easy to recognize that if in (7.15) one identifies Y_r with $Y_r^{(m)}$, V_m coincides with the expression of the isothermal potential elastic energy V.

I shall say that the *sextuple* $Y_r^{*\,(m)}$ *represents a polynomial approximation of order m to the solution of the elastic equilibrium problem if, and only if, it is expressed by*

$$Y_r^{*\,(m)} = \sum_{t=0}^{m} \frac{\overline{Y_r^{*\,(m)} P_t}}{\varrho_t^2} P_t , \tag{7.18}$$

where m + 1 sextuples $\overline{Y_r^{*\,(m)} P_t}$ *belong to the group* G_m *which makes* V_m *minimum in the set* I_m. This is so for the following two reasons:

a) $Y_r^{*\,(m)}$ satisfy equations

$$\int_{C^*} (Y_{rs}^{,s} - k^* F_r) P_t dC^* + \int_{\Sigma^*} (Y_{rs} N_*^s - f_r^*) P_t d\Sigma^* = 0, \quad (t = 0, 1, \cdots, m), \tag{7.19}$$

as is evident since $\overline{Y_r^{(m)} P_t}$ are of class M.

b) $Y_r^{*(m)}$ satisfy SAINT-VENANT's integrability conditions, so that (7.18) may be interpreted as stress components, since a displacement corresponding to them exists.

To justify reason b), let us denote by S_m the set of external forces corresponding to stress $Y_r = Y_r^{*(m)}$, supposing that the sextuple $\overline{Y_r^{*(m)} P_t}$ possesses the above minimum property. Any solution of the set of equations

$$
\left.
\begin{aligned}
Y'^s_{rs} &= k^* \Psi_r^{(m)} \qquad \text{(in } C^*) , \\
Y'_{rs} N^s_* &= \varphi_r^{(m)} \qquad \text{(on } \Sigma^*) ,
\end{aligned}
\right\}
\tag{7.20}
$$

where $k^*\Psi_r^{(m)}, \varphi_r^{(m)}$ are the external forces of S_m, is expressible in the form

$$
Y_r = Y_r^{*(m)} + Y'_r , \tag{7.21}
$$

obtained by associating to the sextuple $Y_r^{*(m)}$ an arbitrary one which is a solution of the homogeneous equations corresponding to (7.20). The Y'_r are expressible by a series like (7.13) and may be represented as a sum of two terms,

$$
Y'_r = Y_r^{'(m)} + Y''_r , \tag{7.22}
$$

where $Y_r^{'(m)}$ are expressions like (7.16), consisting in the first $m + 1$ terms of the series (7.13) which expresses Y'_r. Since it is evident that if the elements of a group G_m of I_m are added to their counterparts $\overline{Y_r^{'(m)} P_t}$ when the external forces are zero one still obtains a group G_m of I_m, it is clear that the sextuple $Y_r^{*(m)} + Y_r^{'(m)}$ is formed by groups G_m of I_m. Further, such sextuples have their elements orthogonal to the elements of the sextuples Y''_r. By orthogonality between Y''_r and $Y_r^{*(m)} + Y_r^{'(m)}$, one deduces that the potential energy corresponding to Y_r expressed by (7.21) is

$$
V = V'_m + \frac{1}{2} \sum_{r,s=1}^{6} \int_{C^*} m_{rs} Y''_r Y''_s \, dC^* , \tag{7.23}
$$

where the expression for V'_m is obtained from (7.17) by substituting $Y_r^{*(m)} + Y_r^{'(m)}$ for $Y_r^{(m)}$ in the second member. By Menabrea's theorem the solution is represented by the sextuple $Y_r^{*(m)} + Y'_r$ which makes V minimum when Y'_r varies in the class of the solutions of homogeneous equations corresponding to (7.20); that is, when G_m varies in I_m and Y''_r varies in the class of the solutions of homogeneous equations corresponding to (7.20) which are orthogonal to $P_0, P_1, \cdots, P_m$.

It follows that the Y'_r are all equal to zero. Therefore, the $Y_r^{*(m)}$ satisfy Menabrea's theorem corresponding to a particular set of external forces and therefore derive from a real displacement (LOCATELLI; see also FINZI, L.).

§ 4. Integration method

Fundamental for the integration of equations (7.1), (7.2), (7.5) is the following property: *the functions*

$$Y_r = \lim_{m \to \infty} Y_r^{(m)}, \tag{7.24}$$

if they exist, satisfy almost everywhere equations (7.1) and equations (7.2), where the external forces are specified if, and only if, the sextuples $\overline{Y_r^{(m)} P_t}$ *belong to the same group* G_m. This property follows from the fact that equations (6.6), considered for all elements of the sequence $\{w_t\}$, are equivalent to the equations

$$\int_{C^*} (Y'_{rs} - k^* F_r)\, P_t dC^* + \int_{\Sigma^*} (Y_{rs} N^s_* - f^*_r)\, P_t d\Sigma^* = 0 \qquad (t=0,1\cdots). \tag{7.25}$$

Then, assuming that Ψ and φ [see (7.11)], respectively, are equal to the expressions which multiply P_t in the integral of (7.25), and keeping in mind that when (7.11) is satisfied for any value of t, Ψ is almost everywhere zero in C^* and φ on Σ^*, we deduce that the basic equations of static elasticity (7.1), (7.2) for the determination of the square-integrable solutions are equivalent to the set of integral equations (7.25) [or (6.6)]. The above property then follows from the fact that the Y_r defined by (7.24) satisfy (7.25) for any value of t.

From what has been said it follows that any solution of (7.1), (7.2) is expressible by means of (7.16), (7.24). The potential energy V corresponding to expressions (7.24) of Y_r is just

$$V = \lim_{m \to \infty} V_m, \tag{7.26}$$

where V_m is expressed by (7.17). This follows from a well known property (PICONE [3]) of the integral product of functions expanded in series (7.24). Let V^*_m be the potential energy corresponding to sextuple $Y^{*\,(m)}_r$, according to (7.17), (7.18). Evidently,

$$V^*_m < V_m \tag{7.27}$$

for any sextuple $Y_r^{(m)}$ different from $Y^{*\,(m)}_r$. Remembering (7.17), one recognizes that the sequence of the minimum $\{V^*_m\}$ is not decreasing. The convergence of the sequence $\{V^*_m\}$ follows from (7.26), (7.27). Consequently any $Y^{*\,(m)}_r$ converges towards a square-integrable function [remember that $\sum_{r,s=1}^{6} m_{rs} Y^{*\,(m)}_r Y^{*\,(m)}_s$ is a positive-definite form]. Now it is possible to conclude that *the six components* Y^*_r *of stress actually present in the body are expressed by*

$$Y^*_r = \lim_{m \to \infty} Y^{*\,(m)}_r. \tag{7.28}$$

In fact, the potential energy V^* corresponding to Y_r^* coincides, according to (7.26), with

$$V^* = \lim_{m \to \infty} V_m^* , \qquad (7.29)$$

and then [see (7.27)]

$$V^* < V(Y_r) \qquad (7.30)$$

for any sextuple Y_r different from Y_r^* and satisfying the equilibrium equations in C^* and on $\Sigma^* - \Sigma_1^*$.

In other words, Y_r^* defined by (7.28) satisfies (7.1) in C^*, (7.2) on $\Sigma^* - \Sigma_1^*$, and Menabrea's theorem. This is sufficient to demonstrate what has been said about Y_r^*, as expressed by (7.28).

Observation I—The above integration method presupposes a complete characterization of the set of forces which may be exerted by the constraints. If the body is clamped on a portion Σ_1^* of the boundary, the Φ_r which are present in (7.14) must satisfy only the fundamental equations of statics. Then, put

$$b_{\eta \tau \lambda}^{(r)} = b_{\eta \tau \lambda}^{'(r)} + c_{\eta \tau \lambda}^{(r)} , \qquad (7.31)$$

where the $b_{\eta \tau \lambda}^{'(r)}$ are obtained by substituting in (6.3) for the integrals corresponding to Σ^* others corresponding to $\Sigma^* - \Sigma_1^*$, while $c_{\eta \tau \lambda}^{(r)}$ need satisfy only the equations obtained by introducing (7.31) in (6.5).

Instead, if the body is supported without friction on Σ_1^* with bilateral constraint, the vector Φ is perpendicular to the support surface. For example, if Σ_1^* is a portion of plane $y_3 = 0$, then $\Phi_1 = \Phi_2 = 0$ on Σ_1^*, and from (7.14) it follows that only that $c_{\eta \tau 0}^{(3)}$ are generally not zero but must satisfy (6.5).

The question is more difficult if the body is supported without friction by unilateral constraint on Σ_1^*.[1] The above method determines the forces exerted by constraints on Σ_1^* or, practically, their approximating expressions. In the case of a unilateral support constraint on Σ^*, these forces, if not zero, must be perpendicular to the supporting surface and oriented towards the region which contains the body. If the solution does not satisfy this condition on a portion Σ_1' of Σ_1^*, one *must* presume that the body detaches[2] itself from the support on Σ_1'. Then, constraints are to be supposed absent on Σ_1' and surface forces to be null there. This permits the application of the integration method not only to the cases of clamped bodies or bodies supported with bilateral constraints but also to the case of bodies supported with unilateral constraint: in

---

[1] The constraint of unilateral rigid or non-rigid support is explicitly taken into consideration by SIGNORINI [9, 10, 12]. In these papers the analytical behavior of such a constraint is studied.

[2] The body being slightly deformable, one excludes the possibility that some portion of $\Sigma^* - \Sigma_1^*$, which in the unstressed state is without constraint, may come in contact with the support surface.

this case, in the beginning, one presumes that there is contact at all points of the support surface Σ_1^*; then, if the corresponding solution shows that there is detachment on a portion Σ_1' of Σ_1^*, one may again apply the procedure, supposing that the body is supported only on $\Sigma_1^* - \Sigma_1'$, etc.

In such manner the above integration method requires the determination of the minimum of only the elastic potential energy expressed by means of Y_{rs} [see (7.4)]; that is, without the presence of the term which expresses the work done by constraint forces. Further, it is to be remembered that, whatever the constraints, the solution for which the constraint forces belong to a set which the constraints are able to exert must give displacements which satisfy geometrical conditions imposed by the constraints. If it is not so, one must presume that the analytical boundary-value problem considered has no solution.

Observation II—In the case of a dislocation the above integration method is useful for determining the supplementary regular stress which, added to the stress caused by the multi-valued displacement characterizing the dislocation, annuls the external body and surface forces.

§ 5. Validity of Menabrea's theorem in non-isothermal cases

1. It is well known that equations (7.1), (7.2) are valid also if thermal phenomena are present. Then, one must presume that the previous integration method, valid in the isothermal case and in the case when the thermodynamic potential is the sum of a function of the strain only and of a function of the temperature only [ignorable temperature], is applicable to some other cases.

In the isothermal case the temperature is known, and the Saint-Venant integrability equations in stress do not depend on it. The same is true in the case of ignorable temperature. On the other hand in the general case, Saint-Venant's equations, if expressed in stress components, depend on the unknown temperature, and the mechanical problem is to be considered together with the thermal problem. But, if the transition from a natural state to a near equilibrium state is adiabatic or is accompanied by a steady heat flow, the temperature may be supposed known, in the first case, by reason of the constancy of the entropy and, in the second case, by the fact that the thermal problem is independent of the mechanical one and may be solved separately. One may extend MENABREA's theorem to these cases and adapt to them the integration method of the previous paragraph (GRIOLI [6]).

2. *Steady case.* As is well known [(4.11, 1)], in the non-isothermal linear case expressions of Y_r are

$$Y_r = - \sum_{s=1}^{6} M_{rs}\varepsilon_s + L_r\tau \qquad (r = 1, 2, \cdots, 6), \qquad (7.32)$$

where the L_r are the thermal coefficients of tension and τ is the temperature variation. Denoting by m_{rs} the quotient obtained by dividing the cofactor of M_{rs} in the determinant $\|M_{rs}\|$ by the determinant [coefficients of $W(Y)$] and putting

$$\eta_r = Y_r - L_r \tau \qquad (r = 1, 2, \cdots, 6), \qquad (7.33)$$

from (7.32) follows

$$\varepsilon_r = - \sum_{s=1}^{6} m_{rs} \eta_s \qquad (r = 1, 2, \cdots, 6). \qquad (7.34)$$

The ε_r have expressions analogous to those in the isothermal case. That is,

$$\varepsilon_r = - \frac{\partial W(\eta)}{\partial \eta_r} \qquad (r = 1, 2, \cdots, 6), \qquad (7.35)$$

where $W(\eta)$ is the expression obtained from the isothermal elastic potential energy density by substituting η_r for Y_r. Then it is sufficient to recall that the proof of Menabrea's theorem is based substantially on the validity of an equality of the kind (7.35) and that in the steady case the temperature is a known function [previously determined] of coordinates, to deduce its validity in the steady case, if $W(Y)$ is replaced by $W(\eta)$. Therefore putting

$$V^{(s)} = \int_{C^*} W(\eta) \, dC^* \qquad (7.36)$$

and supposing τ known, one has: *in the case of steady heat flow, among all stresses which satisfy equations* (7.1) *and the boundary conditions where the forces are specified, the real solution is the one which minimizes the difference between $V^{(s)}$ and the potential of the constraint forces.*

The sextuple Y_r satisfying Menabrea's theorem satisfies the integrability conditions deduced from those of Saint-Venant by substituting for the ε_r their expressions (7.35).

3. *Adiabatic case.* In an adiabatic transformation the entropy does not vary; then

$$\tau = - \frac{1}{c} \sum_{s=1}^{6} L_s \varepsilon_s, \qquad (7.37)$$

where c has the same meaning as in (6.130).

Denoting by

$$M_{rs}^{(a)} = M_{rs} + \frac{L_r L_s}{c} \qquad (r, s = 1, 2 \cdots, 6) \qquad (7.38)$$

the *adiabatic elastic coefficients* and putting

$$W^{(a)}(\varepsilon) = \frac{1}{2} \sum_{r,s=1}^{6} M_{rs}^{(a)} \varepsilon_r \varepsilon_s , \qquad (7.39)$$

from (7.32), (7.37) follows

$$Y_r = -\frac{\partial W^{(a)}(\varepsilon)}{\partial \varepsilon_r} \qquad (r = 1, 2 \cdots, 6) . \qquad (7.40)$$

If $m_{rs}^{(a)}$ denotes the quotient of the cofactor of $M_{rs}^{(a)}$ in the determinant $\|M_{rs}^{(a)}\|$ divided by the determinant, from (7.40) we have

$$\varepsilon_r = -\frac{\partial W_*^{(a)}(Y)}{\partial Y_r} \qquad (r = 1, \cdots 6) , \qquad (7.41)$$

where

$$W_*^{(a)}(Y) = \frac{1}{2} \sum_{r,s=1}^{6} m_{rs}^{(a)} Y_r Y_s . \qquad (7.42)$$

Putting

$$V^{(a)} = \int_{C^*} W_*^{(a)}(Y) dC^* , \qquad (7.43)$$

one recognizes that Menabrea's theorem is valid with respect to the elastic potential energy $V^{(a)}$ and that the minimizing sextuple Y_r [according to Menabrea's theorem] satisfies the integrability equations obtained from those of Saint-Venant by substituting the second members of (7.41) for ε_r.

§ 6. Adaptation of the integration method of § 4 to non-isothermal cases

The validity of Menabrea's theorem in the case of steady heat flow and in the adiabatic one allows us to adapt to these cases the integration method of the linear problem of isothermal elastic statics (GRIOLI [6]). To this end it is sufficient to choose among all solutions of the equilibrium equations, characterized by the integral equality (6.6), the one which satisfies Menabrea's theorem in the steady or adiabatic case. Therefore, the polynomial expansions (7.16), (7.24) can be retained, and the solution is still characterized by (7.18), (7.28) if for V is substituted, respectively, $V^{(s)}$ in the steady case or $V^{(a)}$ in the adiabatic one. In the latter case $Y_r^{*(m)}$ satisfies the integrability conditions for any value of m. In the case of steady heat flow one must assume for τ a power series in P_t, certainly allowed if τ is square-integrable, and assume as an approximate expression of order m

$$\tau^{(m)} = \sum_{t=0}^{m} \frac{\overline{\tau P_t}}{\varrho_t^2} P_t . \qquad (7.44)$$

The expression of $V_m^{(s)}$ which corresponds to (7.17) is

$$V_m^{(s)} = \frac{1}{2} \sum_{t=0}^{m} \frac{1}{\varrho_t^2} \sum_{r,s=1}^{6} m_{rs} \, (\overline{Y_r P_t} - L_r \overline{\tau P_t}) \, (\overline{Y_s P_t} - L_s \overline{\tau P_t}) \, . \qquad (7.45)$$

Then the $Y_r^{*\,(m)}$ which minimize $V_m^{(s)}$ do not satisfy the integrability conditions with reference to real temperature but to its approximate expression of order m given by (7.44).

§ 7. On the integration of the basic problem of isothermal elastic statics in the case of finite deformation

1. The integration of the fundamental problem of isothermal elastic statics in the case of large strain is very difficult because of its non-linearity. In this field a method due to RIVLIN [1] is interesting: the inverse method. RIVLIN supposes certain kinds of deformations and looks for the forces which are able to produce them. Results are found concerning, in general, incompressible bodies [pure extension, hydrostatic pressure, simple extension, simple shear, etc.].

Also interesting in this field are the results of SIGNORINI [4, 11] who, using the inverse method, has studied, with body forces absent, the problem of bending and traction of a body which in the unstressed state has the shape of a rectangular plate with constant thickness which in the stressed state becomes a cylindrical shell sector. More recent results which in general concern particular problems of finite deformation have been obtained by several authors, as for example RIVLIN and SAUNDERS, GREEN and SPRATT, GREEN [1], ADKINS, MANACORDA [2], BLACKBURN and GREEN, etc.

In the present section I shall show that it is possible to establish a general method of integration also in the case of large strain, based on the integral properties used in the linear case.

2. In Chapter IV it has been seen that if the vector u expressing the elastic displacement may be expanded in a power series in a parameter θ,

$$u_r = \sum_{i=1}^{\infty} u_r^{(i)} \, \frac{\theta^i}{i!} \, , \qquad (7.46)$$

analogous developments may be written for ε_r, Y_r, etc. The general terms of these power series for any value of i satisfy differential equations of the same kind as in the linear case, but the body and surface forces are to be replaced by special expressions formed by means of the solutions of the previous sets of equations.

Specifically, if C^* is a natural equilibrium state $[Y_{rs}^{(0)} = 0]$, the basic set of index i is $[(4.18), (4.34)]$:

$$Y_{rs}^{(i),s} = h^* F_r^{(i)} \qquad \text{(in } C^*) \, ,$$

$$Y_{rs}^{(i)} N_*^s = \left\langle \begin{matrix} f_r^{(i)} + \Phi_r^{(i)} & \text{(on } \Sigma_1^*) \, , \\ f_r^{(i)} & \text{(on } \Sigma_2^*) \, , \end{matrix} \right. \qquad (7.47)$$

where $F_r^{(i)}, f_r^{(i)}$ are the applied external forces if $i = 1$, while they are expressed by (4.19) if $i > 1$. With (7.47) are to be associated (4.6).

In Chapter V theorems of existence and uniqueness, valid when constraints are absent $[\Sigma_1^* = 0]$, have been given for the solutions of the basic equations of isothermal elastic statics, and also special cases have been indicated for which an expansion of kind (7.46) does not exist.

Clearly the integration method valid in the linear case must be adaptable to the successive sets of equations (7.47), if Menabrea's theorem may be suitably extended. Naturally, one must suppose that the infinitesimal rotations have been determined corresponding to each term, so that the necessary conditions (4.35) are satisfied.

3. In Chapter IV [§ 1] it has been shown that $Y_r^{(n)}$ is the sum of a linear form in $\varepsilon_i^{(n)}$ with coefficients independent of n and of a polynomial of degree n in $\varepsilon_i^{(1)}, \varepsilon_i^{(2)}, \cdots \varepsilon_i^{(n-1)}$. On the other hand, according to $(1.11), (1.12)$ one has

$$\varepsilon_{rs}^{(n)} = e_{rs}^{(n)} + l_{rs}^{(n)} \, , \qquad e_{rs}^{(n)} = \frac{1}{2} \left(u_{r,s}^{(n)} + u_{s,r}^{(n)} \right) , \qquad (7.48)$$

where the $l_{rs}^{(n)}$ $(n = 1, 2, \cdots,)$ depend only on $u_i^{(1)}, \cdots u_i^{(n-1)}$. Then, putting

$$M_{rs} = \left(\frac{\partial^2 J}{\partial \varepsilon_r^{(1)} \, \partial \varepsilon_s^{(1)}} \right)^{(0)} , \qquad (7.49)$$

from (4.12) we see that

$$Y_r^{(n)} = \eta_r^{(n)} + q_r^{(n)} \qquad (n = 1, 2, \cdots) \, , \qquad (7.50)$$

with

$$\eta_r^{(n)} = - \sum_{s=1}^{6} M_{rs} e_s^{(n)} = - \frac{\partial W (e^{(n)})}{\partial e_s^{(n)}} \qquad (n = 1, 2, \cdots) . \qquad (7.51)$$

In (7.51)

$$W (e^{(n)}) = \frac{1}{2} \sum_{r,s=1}^{6} M_{rs} e_r^{(n)} e_s^{(n)} \qquad (n = 1, 2, \cdots) \qquad (7.52)$$

is the expression obtained from the isothermal elastic potential energy by substituting $e_r^{(n)}$ for ε_r, while the $q_r^{(n)}$ are constructed by means of $l_r^{(n)}$ and of terms not explicitly written in (4.12) and are known functions of

$u_r^{(1)}, \cdots, u_r^{(n-1)}$. For example, in the isotropic case we have

$$W(e^{(n)}) = \frac{1}{2}\left[(\lambda + 2\mu)\, I_1^2(e^{(n)}) - \mu\, I_2(e^{(n)})\right].\tag{7.53}$$

From what has been said, it follows that the set of equations of index n is[1]

$$\eta_{rs}^{(n)\,'\,s} = k^* \Psi_r^{(n)}\qquad\qquad\text{(in } C^*\text{)},$$
$$\eta_{rs}^{(n)} N_*^s = \begin{cases}\varphi_r^{(n)} + \varPhi_r^{(n)} & \text{(on } \Sigma_1^*\text{)},\\[4pt]\varphi_r^{(n)} & \text{(on } \Sigma_2^*\text{)},\end{cases}\tag{7.54}$$

where [see (4.19)]

$$k^* \Psi_r^{(n)} = k^* F_r^{(n)} - q_{rs}^{(n),\,s}, \qquad \varphi_r^{(n)} = f_r^{(n)} - q_{rs}^{(n)} N_*^s \qquad (n = 1, 2, \cdots).\tag{7.55}$$

According to (7.55), the second members of (7.54) are known functions of $u_r^{(1)}, \cdots u_r^{(n-1)}$ and $\varPhi_r^{(n)}$ and satisfy the necessary integrability conditions of the successive sets of equations if $k^* F_r^{(n)}, f_r^{(n)}, \varPhi_r^{(n)}$ satisfy those conditions.

4. MENABREA's theorem holds for equations (7.54) if one interprets $\eta_{rs}^{(n)}$ as stress components, $k^* \Psi_r^{(n)}$ as body forces and $\varphi_r^{(n)}, \varPhi_r^{(n)}$ as surface forces. To justify this fact, first I observe that, putting

$$W(\eta^{(n)}) = \frac{1}{2}\sum_{r,s=1}^{6} m_{rs}\,\eta_r^{(n)}\eta_s^{(n)}\tag{7.56}$$

where m_{rs} are the quotients between the cofactors of M_{rs} in the matrix $|M_{rs}|$ and their determinant $\|M_{rs}\|$, one has

$$e_r^{(n)} = -\frac{\partial W(\eta^{(n)})}{\partial \eta_r^{(n)}}\qquad (n = 1, 2, 3, \cdots),\tag{7.57}$$

analogous to the formula of linear theory.

Let $\delta\eta_{rs}^{(n)}$ be variations of $\eta_{rs}^{(n)}$ which satisfy the equations

$$\begin{aligned}[\delta\eta_{rs}^{(n)}]^{\cdot\,s} &= 0 & \text{(in } C^*\text{)},\\[4pt]\delta\eta_{rs}^{(n)} N_*^s &= 0 & \text{(on } \Sigma^* - \Sigma_1^*\text{)}.\end{aligned}\qquad\Biggr\}\tag{7.58}$$

From (7.58) follows

$$\int_{C^*} [\delta\eta_{rs}^{(n)}]^{\cdot\,s}\, u^{(n)\,r}\, dC^* + \int_{\Sigma^*-\Sigma_1^*} \delta\eta_{rs}^{(n)}\, u^{(n)\,r} N_*^s\, d\Sigma^* = 0.\tag{7.59}$$

Following the usual method and bearing in mind that $u_r^{(n)} = 0$ for $n > 1$ on Σ_1^*, from (7.59) one deduces: *among all sextuples $\eta_{rs}^{(n)}$ which satisfy (7.54, 1) in C^* and (7.54, 2) on $\Sigma^* - \Sigma_1^*$ the real one is the one*

---

[1] Plainly, it is to be supposed that $\eta_{rr}^{(n)} = \eta_r^{(n)}$ and $\eta_{r+1\,r+2}^{(n)} = \eta_{r+3}^{(n)}$.

which minimizes

$$V(\eta^{(n)}) = \int_{C*} W(\eta^{(n)}) \, dC* - \int_{\Sigma_1^*} \Phi_r^{(n)} u^{(n)r} \, d\Sigma_1^* \qquad (n = 1, 2, \cdots) . \qquad (7.60)$$

For example, if the displacement $s = \theta s'$, with s' independent of θ, is assigned on Σ_1^*, in (7.60) the second integral vanishes for $n \geqslant 2$. Plainly, the same thing happens for any value of n if the body is clamped on Σ_1^*, or if constraints are absent.

Further, following a procedure analogous to that of the linear case, it is possible to show that the $\eta_r^{(n)}$ which minimize $V(\eta^{(n)})$ make $e_r^{(n)}$, as given by (7.57), integrable for the determination of $u_r^{(n)}$.

5. Although the theorems of Chapter V showing that u_r, Y_r may be expressed by power series in θ were established subject only to the absence of constraints, I wish also to consider, more generally, the cases of clamping and supporting constraints, it being my opinion that the possibility of expanding solutions in power series in θ exists also for these cases. A demonstration of this fact is certainly desirable but still unknown.

To show the possibility of applying the integration method of the linear case to each of successive sets of fundamental equations, I shall consider three cases:

a) *Absence of constraints.* $\Phi_r^{(n)}$ are zero and Σ_1^* vanishes. The solution of the problem is represented by

$$\left. \begin{array}{l} \eta_r^{(i)} = \lim_{m \to \infty} \eta_r^{(i)\,(m)} , \\[2mm] \eta_r^{(i)\,(m)} = \sum_{t=0}^{\infty} \frac{1}{\varrho_t^2} \overline{\eta_r^{(i)\,(m)} P_t} \cdot P_t , \end{array} \right\} \qquad (7.61)$$

where $\overline{\eta_r^{(i)\,(m)} P_t}$ minimizes the expression

$$V_m^{(i)} = \frac{1}{2} \sum_{t=0}^{m} \frac{1}{\varrho_t^2} \sum_{r,s=1}^{6} m_{rs} \overline{\eta_r^{(i)\,(m)} P_t} \cdot \overline{\eta_s^{(i)\,(m)} P_t} , \qquad (7.62)$$

keeping in mind, in the sense explained in §§ 3, 4, the integral equations

$$\lambda \overline{\eta_{r1}^{(n)} y_1^{\lambda-1} y_2^{\eta} y_3^{\tau}} + \eta \overline{\eta_{r2}^{(n)} y_1^{\lambda} y_2^{\eta-1} y_3^{\tau}} + \tau \overline{\eta_{r3}^{(n)} y_1^{\lambda} y_2^{\eta} y_3^{\tau-1}} = b_{\lambda\eta\tau}^{(r,n)} \qquad (7.63)$$

where

$$b_{\lambda\eta\tau}^{(r,n)} = -\frac{1}{C*} \left\{ \int_{C*} y_1^{\lambda} y_2^{\eta} y_3^{\lambda} k* \Psi_r^{(n)} \, dC* + \int_{\Sigma^*} y_1^{\lambda} y_2^{\eta} y_3^{\lambda} \varphi_r^{(n)} \, d\Sigma* \right\} . \qquad (7.64)$$

b) *Clamping constraint.* If the body is clamped on a portion Σ_1^*, of its boundary, (7.61), (7.62), (7.63), (7.64) are valid but the minimum of $V_m^{(i)}$ is to be found in a larger class, since now $b_{\lambda\eta\tau}^{(r,n)}$ depend on unknown terms

$c_{\lambda \eta \tau}^{(r,n)}$ referring to $\Phi_r^{(n)}$ and analogous to (7.14). Specifically, we have

$$c_{\lambda \eta \tau}^{(r,n)} = - \frac{1}{C^*} \int_{\Sigma_1^*} y_1^\lambda y_2^\eta y_3^\tau \Phi_r^{(n)} d\Sigma_1^* . \tag{7.65}$$

The $c_{\lambda \eta \tau}^{(r,n)}$ are to be formed by means of $\Phi_r^{(n)}$ which satisfy (4.35), (4.39), (4.40). Then the $c_{\lambda \eta \tau}^{(r,n)}$ are all arbitrary except $c_{000}^{(r,n)}$, which vanishes for $n \geqslant 2$ and $c_{000}^{(r,1)}$, $c_{100}^{(2,n)}$, etc., which must verify (4.35), (4.40), for $n \geqslant 1$.

The integration method is analogous to the one in §§ 3, 4, and I shall not repeat it [see also Observation I, § 4].

c) *Supporting constraint without friction.* If the constraint is bilateral, the case is analogous to the clamping one. For example, if the body is supported on portion Σ_1^* of plane $y_3 = 0$, it is sufficient to put $\Phi_1^{(n)} = \Phi_2^{(n)} = 0$ $(n = 1, 2, \cdots)$, etc.

Instead, if the constraint is unilateral, the analytical conditions which are present are very complicated, since not even the support surface is known. Also if the portion Σ_1^* of the boundary of the natural state where there is contact is known, in the stressed state the contact may be present on a portion Σ' of Σ^* which, generally, includes portions of both Σ_1^* and $\Sigma^* - \Sigma_1^*$. But, if one supposes, as certainly is allowed in some cases, that in the stressed configuration the contact does not take place outside of Σ_1^*, the previous integration method may be applied provided that one keeps in mind that the $\Phi_r^{(n)}$ satisfy certain conditions which derive from support conditions, for any value of n.

For example, if Σ_1^* is a portion of the plane $y_3 = 0$ and the body is contained in semi-space $y_3 \geqslant 0$, the vector Φ^* satisfies the conditions

$$\Phi_1^* = \Phi_2^* = 0 . \tag{7.66}$$

(7.66) certainly are valid for $\Phi_r^{(n)}$. We shall have for any value of n

$$\Phi_1^{(n)} = \Phi_2^{(n)} = 0 \qquad (n = 1, 2, \cdots) . \tag{7.67}$$

Nothing can be said about $\Phi_3^{(n)}$ for $n \geqslant 2$, except that it must satisfy (4.35), (4.39). In conclusion, the method is applicable, but — if a solution exists—it may be possible to determine the real support surface so that

$$\left. \begin{array}{ll} \Phi_3^* \geqslant 0, & u_3 = 0 \\ \Phi_3^* = 0, & u_3 > 0 \end{array} \right\} \quad (\text{on } \Sigma_1^*) , \tag{7.68}$$

where only the upper or lower inequalities are valid.

§ 8. Integration method of M. Picone

1. Interesting for numerical applications is an integration method of the linear elastic static problem proposed by M. PICONE [4, 6]. It operates directly on the components of displacement and is based on

BETTI's reciprocity theorem, the functional interpretation of which allows one not only to establish existence theorems (FICHERA [4]) but also to construct a method of calculation of the solutions of the basic problem of isothermal linear elastic statics.

To begin with, PICONE constructs a sequence of polynomial vectors which are solutions of the homogeneous equations (5.64). Put

$$p'_h(y_1, y_2) + \sqrt{-1}\, p''_h(y_1, y_2) = (y_1 + \sqrt{-1}\, y_2)^h, \left.\begin{array}{c}\\\\\end{array}\right\} \quad (7.69)$$
$$r = \sqrt{y_1^2 + y_2^2 + y_3^2}\,;$$

let us denote by $\zeta_n(\eta)$ the Legendre polynomial in η of degree n. It is easily recognized that the functions

$$v_{n\,2l} = r^{n-l}\zeta_n\left(\frac{y_3}{r}\right)p'_l(y_1, y_2)\,, \quad v_{n\,2l-1} = r^{n-l}\zeta_n\left(\frac{y_3}{r}\right)p''_l(y_1, y_2)$$
$$(l = 0, 1, \cdots n)\,, \quad (7.70)$$

are harmonic, homogeneous polynomials of degree n. Let

$$q = \frac{\lambda}{\mu}\,, \qquad q_n = -\frac{1+q}{2q\,(n-1)+6n-4}\,. \qquad (7.71)$$

Let us denote by $\{V_i\}$ the sequence of vectors $V_{nk}^{(s)}$, $(s = 1, 2, 3; n = 0, 1, \cdots; k = 0, 1, \cdots 2\,n)$ whose components $[V_{nk}^{(s)}]_t$ are expressed by

$$[V_{nk}^{(s)}]_t = \delta_{ts}v_{nk} + q_n r^2 v_{nk,\,ts}\,. \qquad (7.72)$$

It is easily demonstrated that the $3\,(2\,n+1)$ vectors $V_{nk}^{(s)}$ are homogeneous polynomial solutions of degree n of equations (5.64). Further, it is possible to demonstrate that formula (7.72) gives all such linearly independent solutions of equations (5.64), and these solutions are called *homogeneous elasticity polynomials of degree n*.

2. According to PICONE's method, let us consider BETTI's theorem with reference to the displacement u and the vector V_i. With reference to *Problem* 3) [see Chapter V, § 6, n. 2], one has

$$\int_{\Sigma_1^*} u \cdot L(V_i)\,d\Sigma_1^* - \int_{\Sigma_1^*} V_i \cdot L(u)\,d\Sigma_1^* + \int_{\Sigma_2^*} u \cdot L(V_i)\,d\Sigma_2^* -$$
$$-\int_{\Sigma_2^*} V_i \cdot f^*\,d\Sigma_2^* - \int_{C^*} k^* V_i \cdot F^*\,dC^* = 0 \qquad (i = 1, 2, \cdots, n)\,. \qquad (7.73)$$

Put

$$u = \sum_{j=1}^{n} \gamma_j L(V_j) \qquad (\text{on } \Sigma_2^*)\,, \qquad (7.74)$$

$$L(u) = -\sum_{j=1}^{n} \gamma_j V_j \qquad (\text{on } \Sigma_1^*)\,, \qquad (7.75)$$

and, given n, the coefficients γ_j are to be determined by introducing (7.74), (7.75) in (7.73). Then one has the equations

$$\sum_{j=1}^{n} a_{ij}\gamma_j = c_i \qquad (i = 1, 2, \cdots, n) , \qquad (7.76)$$

where

$$\left. \begin{aligned} a_{ij} &= \int_{\Sigma_1^*} V_i \cdot V_j d\Sigma_1^* + \int_{\Sigma_2^*} L(V_i) \cdot L(V_j) d\Sigma_2^* , \\ c_i &= \int_{C^*} V_i \cdot k^* F^* dC^* - \int_{\Sigma_1^*} u \cdot L(V_i) d\Sigma_1^* + \int_{\Sigma_2^*} V_i \cdot f^* d\Sigma_2^* . \end{aligned} \right\} \qquad (7.77)$$

Supposing the determinant $\|a_{ij}\|$ different from zero, let $\gamma_j^{(n)}$ be values of γ_j satisfying (7.76), and let

$$\left. \begin{aligned} u^{(n)} &= \sum_{j=1}^{n} \gamma_j^{(n)} L(V_j) \qquad \text{on } \Sigma_2^* , \\ L^{(n)}(u) &= -\sum_{j=1}^{n} \gamma_j^{(n)} V_j , \qquad \text{on } \Sigma_1^* . \end{aligned} \right\} \qquad (7.78)$$

If the body is clamped on a portion Σ_1^* of Σ^* where constraints are present, G. FICHERA [4] has demonstrated the Hilbertian completeness of the set of vectors having components V_{js} on Σ_1^* and $L(V_j)_s$ on Σ_2^*. Further FICHERA has shown that $\|a_{ij}\|$ is different from zero and he has demonstrated the convergence of $u^{(n)}$ to the displacement u on Σ_2^* and of $L^{(n)}(u)$ to the stress $L(u)$ on Σ_1^*.

Thus the external forces on Σ_1^* and the displacement on Σ_2^* are determined. Then by using vectors of SOMIGLIANA's matrix and well known expressions (see also FICHERA [2]), it is possible to express the displacement u and its derivatives within C^*.

Naturally the above method is also applicable for solving *Problems* 1), 2) [Chapter V § 6]: for *Problem* 1) it is sufficient to suppress in (7.77) the integrals corresponding to Σ_2^*, to identify Σ_1^* with Σ^* and to consider only (7.75). On the other hand, for *Problem* 2) it is sufficient to suppress in (7.77) the integrals corresponding to Σ_1^*, to identify Σ_2^* with Σ^* and to consider only (7.74).

3. The completeness property of the sequence $\{V_i\}$ allows construction of the solutions of Problems 1), 2) without using SOMIGLIANA's matrix. According to PICONE's method, let us suppose that

$$u^{(n)} = \sum_{i=1}^{n} c_i^{(n)} V_i , \qquad (7.79)$$

and let us determine coefficients $c_i^{(n)}$ by the equations

$$\sum_{j=1}^{n} c_j^{(n)} \int_{\Sigma^*} V_i \cdot L(V_j) d\Sigma^* = b_i \qquad (i = 1, 2 \cdots, n) . \qquad (7.80)$$

Supposing body forces absent, one may demonstrate (FICHERA [4]) that the *vector* $u^{(n)}$ *expressed by* (7.79) *converges uniformly in* C^* *to the solution of Problem* 1) *if it is supposed that*

$$b_i = \int_{\Sigma^*} u \cdot L(V_i) \, d\Sigma^* \qquad (i = 1, 2 \cdots, n) . \qquad (7.81)$$

Instead, $u^{(n)}$ *converges uniformly within* C^* *to the solution of Problem* 2) *if it is supposed that*

$$b_i = \int_{\Sigma^*} f^* \cdot V_i \, d\Sigma^* . \qquad (7.82)$$

For solving Problem 3) it is necessary to find a set of vectors w_i complete in the class of biregular solutions of equations (5.64) and zero on Σ_1^*, like the one used by FICHERA [4, Chap. VI]. Then, under the hypothesis that the body is clamped on Σ_1^* [$u \equiv 0$ on Σ_1^*] and that body forces are absent, the solution of Problem 3) is expressible in the form

$$u = \lim_{m \to \infty} \sum_{i=1}^{m} c_i^{(m)} w_i , \qquad (7.83)$$

if $c_i^{(m)}$ are determined by means of equations (7.80) and if it is supposed that $V_i \equiv w_i$ and that b_i are expressed by (7.82).

§ 9. On integration of the static problem of the homogeneous plate of arbitrary thickness

1. Clearly, the integration method proposed in § 4 is applicable also to the equilibrium problems of thin bodies, but their thinness makes it possible to simplify the procedure.

In particular, this is possible in the case of a homogeneous plate of arbitrary thickness, even an anisotropic one, and this case will be briefly considered here. For greater detail see GRIOLI [8].

Let C^* be an elastic plate in the form of a right cylinder of height h and cross section A^*. Choose the origin of the rectangular Cartesian coordinates (y_1, y_2, y_3) in the middle surface with y_3 parallel to the generators of the lateral surface σ^*. Denote by l the line (directrix) which is the intersection of σ^* and the cross-section $y_3 = 0$. Let the load density on the base $y_3 = -\frac{1}{2} h$ be $q\,(y_1, y_2)$, directed parallel to y_3. Let σ' be the portion of σ^* subjected to the clamping constraint, l' the corresponding portion of l; let the components of the clamping constraint forces be Φ_1, Φ_2, Φ_3. Let the plate be supported without friction on a portion A' of its lower base π^*; let the components of this supporting force be Ψ_1, Ψ_2, Ψ_3, where $\Psi_1 = \Psi_2 = 0$ and $\Psi_3 \leqslant 0$.

The basic equilibrium equations are

$$Y_{rs}^{;s} = 0 \qquad \text{(in } C^*\text{)}, \tag{7.84}$$

$$Y_{13} = Y_{23} = 0 \qquad \text{for } y_3 = \pm \frac{h}{2},$$

$$Y_{33} = \begin{cases} 0 & \text{for } y_3 = \frac{h}{2} \qquad \text{(in } \pi^* - A') \\ -\Psi_3 & \text{for } y_3 = \frac{h}{2} \qquad \text{(in } A') \\ q & \text{for } y_3 = -\frac{h}{2}. \end{cases} \right\} \tag{7.85}$$

$$Y_{rs}N_*^s = \begin{cases} f_r^* & \text{(on } \sigma^* - \sigma'\text{)}, \\ \Phi_r^* & \text{(on } \sigma'\text{)}. \end{cases} \right\} \tag{7.86}$$

2. Let us consider a sequence $\{Q_t\}$ of polynomials

$$Q_t(y_3) = \frac{d^t}{dy_3^t}\left(y_3^2 - \frac{h^2}{4}\right)^t \qquad (t = 0, 1 \cdots), \tag{7.87}$$

which are orthogonal in the interval $-\frac{h}{2} \leftrightarrow \frac{h}{2}$ and coincide with the Legendre polynomials if coefficients independent of y_3 are neglected.

The components of stress are expressible in this interval by power series in Q_t with coefficients depending on y_1, y_2. Therefore, we may put

$$Y_{rs} = \sum_{i=0}^{\infty} \frac{\xi_{rst}}{\mu_i^2} Q_t, \tag{7.88}$$

where

$$\xi_{rst} = \int_{-\frac{h}{2}}^{\frac{h}{2}} Y_{rs} Q_t dy_3, \left. \begin{array}{c} \\ \\ \\ \\ \end{array} \right\} \tag{7.89}$$

$$\mu_i^2 = \int_{-\frac{h}{2}}^{\frac{h}{2}} Q_i^2 dy_3.$$

$\xi_{rst}(y_1, y_2)$ are defined in the domain A^*.

If $\{v_\lambda\}$ is the sequence of monomials $y_1^\alpha y_2^\beta$ and $\{P_\lambda\}$ that of the polynomials obtained by adding to v_λ that linear combination of $v_0, v_1, \cdots$, $v_{\lambda-1}$ which makes it orthogonal in A^* to those monomials, ξ_{rst} are expressible in A^* by series of P_λ.

Then, putting

$$\varrho_\lambda^2 = \frac{1}{A^*} \int_{A^*} P_\lambda^2 dA^*, \tag{7.90}$$

one has

$$\xi_{rst} = \sum_{\lambda=0}^{\infty} \frac{\overline{\xi_{rst}\, P_\lambda}}{\varrho_\lambda^2}\, P_\lambda \qquad (t = 0, 1, \cdots)\, , \tag{7.91}$$

where the bar denotes the average value in $A*$.

The expression of the elastic potential energy corresponding to (7.88), (7.91) is[1]

$$V = A* \sum_{t=0}^{\infty} \sum_{\lambda=0}^{\infty} \frac{1}{\mu_t^2\, \varrho_\lambda^2} \sum_{r,s=1}^{6} m_{rs}\, \overline{\xi_{rt}\, P_\lambda}\, \overline{\xi_{st}\, P_\lambda}\, . \tag{7.92}$$

3. Denoting by A'' the projection on the plane $y_3 = 0$ of the support surface A', let

$$\beta_t = \begin{cases} q Q_t\left(-\dfrac{h}{2}\right) & \text{(in } A* - A'')\, , \\[2mm] q Q_t\left(-\dfrac{h}{2}\right) + \Psi_3 Q_t\left(\dfrac{h}{2}\right) & \text{(in } A'')\, . \end{cases} \tag{7.93}$$

One may always write [see (7.87)]

$$Q_t = \sum_p{}^* \alpha_{pt}\, y_3^p\, , \qquad Q_{t,3} = \sum_p{}^* p\alpha_{pt}\, y_3^{p-1}\, , \tag{7.94}$$

where the sums are to be taken over $0, 2, 4, \cdots t$ when t is an even number, or $1, 3, 5, \cdots t$, in the contrary case. Putting

$$\eta_{rp} = \int_{-\frac{h}{2}}^{\frac{h}{2}} Y_{r3}\, y_3^p\, dy_3 \qquad (p = 0, 1, \cdots)\, , \tag{7.95}$$

from equations (7.84), (7.85), (7.94) follows

$$\left. \begin{array}{l} \displaystyle \sum_{s=1}^{2} \xi_{rst,s} - \sum_p{}^* p\alpha_{pt}\eta_{r\,p-1} = 0 \qquad (r = 1,2;\, t = 0,1,2,\cdots) \\[4mm] \displaystyle \sum_p{}^* \alpha_{pt}\Big[\sum_{s=1}^{2} \eta_{s\,p,s} - p\,\eta_{3\,p-1}\Big] = \beta_t \qquad (t = 0, 1, 2, \cdots)\, . \end{array} \right\} \tag{7.96}$$

---

[1] Plainly, it is to be supposed that $\xi_{rrt} = \xi_{rt},\ \xi_{r+1\,r+2t} = \xi_{r+3t}$.

The corresponding boundary conditions, obtained from (7.86), are

$$\sum_{s=1}^{2} \xi_{rst} N_*^s = \left\{ \begin{array}{ll} \int_{-\frac{h}{2}}^{\frac{h}{2}} f_r^* Q_t \, dy_3 & (\text{on } l - l'), \\ & (r = 1, 2), \\ \int_{\frac{h}{2}} \varPhi_r^* Q_t \, dy_3 & (\text{on } l'), \end{array} \right.$$

$$\sum_{p}^{*} \alpha_{pt} \sum_{s=1}^{2} \eta_{sp} N_s^* = \left\{ \begin{array}{ll} \int_{-\frac{h}{2}}^{\frac{h}{2}} f_3^* Q_t \, dy_3 & (\text{on } l - l'), \\ \int_{-\frac{h}{2}}^{\frac{h}{2}} \varPhi_3^* Q_t \, dy_3 & (\text{on } l'). \end{array} \right. \tag{7.97}$$

Putting

$$\gamma_{\eta \tau t}^{(r)} = -\frac{1}{A^*} \left\{ \int_{l'} y_1^\eta y_2^\tau \left[\int_{-\frac{h}{2}}^{\frac{h}{2}} \varPhi_r^* Q_t \, dy_3 \right] dl + \right.$$

$$\left. + \int_{l-l'} y_1^\eta y_2^\tau \left[\int_{-\frac{h}{2}}^{\frac{h}{2}} f_r^* Q_t \, dy_3 \right] dl + \delta_{3r} \int_{A^*} \beta_t y_1^\eta y_2^\tau \, dA^* \right\}, \tag{7.98}$$

from (7.96), (7.97), (7.98) follows

$$\left. \begin{array}{l} \eta \overline{\xi_{r1t} y_1^{\eta-1} y_2^\tau} + \tau \overline{\xi_{r2t} y_1^\eta y_2^{\tau-1}} + \sum_{p}^{*} \alpha_{pt} \overline{\eta_{rp-1} y_1^\eta y_2^\tau} = \gamma_{\eta \tau t}^{(r)} \\ \qquad\qquad (r = 1, 2; t = 0, 1 \cdots), \\ \sum_{p}^{*} \alpha_{pt-1} \left[\eta \overline{\eta_{1p} y_1^{\eta-1} y_2^\tau} + \tau \overline{\eta_{2p} y_1^\eta y_2^{\tau-1}} + p \overline{\eta_{sp-1} y_1^\eta y_2^\tau} \right] = \gamma_{\eta \tau t-1}^{(3)}, \\ \qquad\qquad (t = 1, 2, \cdots). \end{array} \right\} \tag{7.99}$$

4. The basic equations of statics are expressed by

$$\left. \begin{array}{ll} \gamma_{000}^{(r)} = 0 & (r = 1, 2, 3), \\ \gamma_{100}^{(2)} - \gamma_{010}^{(1)} = \gamma_{100}^{(3)} - \gamma_{001}^{(1)} = \gamma_{001}^{(2)} - \gamma_{010}^{(3)} = 0. \end{array} \right\} \tag{7.100}$$

If $\varrho_{\eta \tau t}^{(r)}$ is the unknown term contained in the expression of $\gamma_{\eta \tau t}^{(r)}$, to each choice of $\overline{\xi_{rst} y_1^\eta y_2^\tau}$, $\overline{\eta_{rp} y_1^\eta y_2^\tau}$ and $\varrho_{\eta \tau t}^{(r)}$ which satisfies (7.99), (7.100) there corresponds a group of $\overline{\xi_{rst} P_\lambda}$ ($r, s = 1, 2, 3$).

For any value of t, let I_{mt} be the set of all sextuples of $\overline{\xi_{rst} P_\lambda}$ ($r, s = 1, 2, 3$) so constructed for $\lambda = 0, 1, \cdots, m$. Considerations analogous to those of §§ 3, 4 assure that each square-integrable solution of the boundary problem (7.96), (7.97) may be expanded in a series (7.91), provided the sextuples $\overline{\xi_{rst}} P_\lambda$ belong to the set I_{mt} for any t. With the ξ_{rst} determined, (7.88) gives all square-integrable solutions of the basic equations (7.84), (7.85), (7.86). Which among these represents the real stress is to be determined by MENABREA's theorem: the series (7.88), (7.91) are to be constructed by using those sextuples $\overline{\xi_{rst}} P_\lambda$ which belong to I_{mt} and minimize the expression (7.92) of V.

5. In the pratical applications of the above method one considers only a few terms of the series (7.88), (7.91), (7.92). The corresponding ploynomials verify the Saint-Venant integrability conditions. The great advantage of expansions of the type (7.88) for studying problems of this kind lies in the fact that since the thickness is small in comparison with the transverse dimension, only a few terms need be taken. The usual plate theory represents merely a first approximation, corresponding to taking only the first two terms in (7.88) for r, $s = 1, 2$. It is possible to show that to this order of approximation the basic formulas of the usual theory of thin plates are obtained. Put

$$N_{rs} = \int_{-\frac{h}{2}}^{\frac{h}{2}} Y_{rs}\, dy_3\,, \qquad M_{rs} = \int_{-\frac{h}{2}}^{\frac{h}{2}} Y_{rs}\, y_3\, dy_3 \qquad (r, s = 1, 2, 3)\,. \qquad (7.101)$$

N_{rr} $(r = 1, 2)$ are the normal stress resultants, N_{13}, N_{23} and N_{12} the shearing ones, M_{12} and $-M_{12}$ the twisting moments, M_{11} and $-M_{22}$ the bending ones. These quantities represent the fundamental elements of the usual theory of thin plates.

From (7.87), (7.89, 1), (7.95), (7.101) follows

$$\begin{aligned} \xi_{rs0} &= N_{rs}\,, \qquad \xi_{rs1} = 2M_{rs} \qquad &(r, s = 1, 2, 3)\,, \\ \eta_{r0} &= \xi_{r30} \qquad &(r = 1, 2)\,, \end{aligned} \qquad \biggr\} \qquad (7.102)$$

which, keeping in mind (7.99), allows us to express the corresponding integral properties for N_{rs}, M_{rs}. Further, taking in (7.92) only the terms corresponding to values 0, 1 of index t, and keeping in mind (7.102), one obtains the expression for the elastic potential energy V which corresponds to the order of approximation of the usual theory of thin plates. For brevity, the reader is referred to the original paper (GRIOLI [8]) for further details. Here I observe only that for a general anisotropic plate the analytical problem depends simultaneously on all the unknowns, while in the isotropic case and in the one of anisotropic bodies for which

$$m_{14} = m_{15} = m_{24} = m_{25} = m_{46} = m_{56} = 0 \qquad (7.103)$$

(that is, for monoclinic, rhombic, hexagonal, tetragonal, or monometric crystals), there are two distinct problems, one for the determination of N_{11}, N_{22}, N_{12} [plane problem] and the other for M_{11}, M_{22}, M_{12}, N_{13}, N_{23} [bending problem].

6. As a test for the method above I have considered a case in which the solution is well known: the bending of a thin square plate of side $2a$, everywhere supported and subjected to a uniform load q, considering the polynomials P_λ up to and including the fourth degree.

Correspondingly, the expressions for M_{11}, etc., are of fourth degree, while the component u_3 of displacement corresponding to the y_3 axis is of sixth degree. The calculations are simple, and the results obtained by means of polynomials of such low degree are sufficiently good.

Specifically, for the maximum of the bending moment and u_3 one finds

$$\left. \begin{aligned} M_{11}(0, 0) = M_{22}(0, 0) &= -0.153(1+\nu)qa^2, \\ u_3(0, 0) &= 0.0652\,a^4\,\frac{12(1-\nu^2)q}{Eh^3} \end{aligned} \right\} \qquad (7.104)$$

[E is Young's modulus and ν is Poisson's coefficient], where according to NADAI [pp. 118, 127]

$$\left. \begin{aligned} M_{11}(0, 0) = M_{22}(0, 0) &= -0.160(1+\nu)qa^2, \\ u_3(0, 0) &= 0.649\,a^4\,\frac{12(1-\nu^2)q}{Eh^3}. \end{aligned} \right\} \qquad (7.105)$$

The constraint conditions according to the usual theory are $M_{ii} = 0$ $(i = 1, 2)$; $u_3 = 0$ at all points of the boundary of A^*. In the above solution, M_{ii} does not vanish identically on the contour of the plate, but its absolute value there has a maximum which is one seventh of its value at the center of the plate. On the other hand, u_3 vanishes identically on the contour of the plate, *even though such a condition has not been directly imposed.*

§ 10. Saint-Venant's problem

Another case where it is convenient to transform the integration method of §§ 3, 4 is the one concerning SAINT-VENANT's problem of cylindrical bodies subjected to external forces acting only upon the two bases.

We do this with the aim of using the approximation hypothesis of SAINT-VENANT's theory in order to simplify the procedure. So doing, one will see that it is necessary to replace MENABREA's theorem by a special minimum property (GRIOLI [7]). If the axis y_3 of the central Cartesian co-ordinate system is parallel to the generators of the cylindrical region occupied by the solid, and if n_1, n_2, $n_3 = 0$ are the direction cosines of the inward normal to the lateral surface σ, according to SAINT-VENANT's theory the basic conditions for stress are the following:

$$Y_{11} = Y_{22} = Y_{12} = 0 \qquad \text{(in } C^*\text{)}, \tag{7.106}$$

$$Y_{13}\, n_1 + Y_{23}\, n_2 = 0 \qquad \text{(on } \sigma\text{)}, \tag{7.107}$$

$$\overline{Y}_{r3} = -\frac{R_r}{A^*} \qquad (r = 1, 2, 3), \tag{7.108}$$

$$\left.\begin{array}{l} \overline{y_2 Y_{33}} - \overline{y_3 Y_{23}} = -\dfrac{M_1}{A^*}, \\[2mm] \overline{y_1 Y_{33}} - \overline{y_3 Y_{13}} = \dfrac{M_2}{A^*}, \\[2mm] \overline{y_1 Y_{23}} - \overline{y_2 Y_{13}} = -\dfrac{M_3}{A^*}, \end{array}\right\} \tag{7.109}$$

where the bar denotes the average value over the cross section A^* of the cylinder. Supposing that the cylinder's height is $2h$ in (7.108), (7.109), R_r, M_r are the components of the resultant and resultant moment of the external forces acting upon the base $y_3 = h$. The three equilibrium equations imply the independence of Y_{13}, Y_{23} from y_3 and the expression

$$Y_{33} = Y'_{33}\, y_3 + Y''_{33}, \tag{7.110}$$

with Y'_{33}, Y''_{33} also being independent of y_3. Then from (7.109) follow

$$\overline{Y}_{r3} = -\frac{R_r}{A^*} \qquad (r = 1, 2), \tag{7.111}$$

$$\overline{Y'}_{33} = 0, \qquad \overline{Y''}_{33} = -\frac{R_3}{A^*}, \tag{7.112}$$

$$\overline{y_r Y'_{33}} = -\frac{R_r}{A^*} \qquad (r = 1, 2), \tag{7.113}$$

$$\overline{y_2 Y''_{33}} = -\frac{M_1}{A^*}, \qquad \overline{y_1 Y''_{33}} = \frac{M_2}{A^*}, \tag{7.114}$$

$$\overline{y_1 Y_{23}} - \overline{y_2 Y_{13}} = -\frac{M_3}{A^*}, \tag{7.115}$$

while of the equilibrium equations only

$$Y_{13,1} + Y_{23,2} + Y'_{33} = 0 \tag{7.116}$$

is to be considered.

The problem is that of finding four functions Y_{13}, Y_{23}, Y'_{33}, Y''_{33} which satisfy equations (7.107), (7.116), (7.112), (7.113), (7.114), (7.115) and derive from a real displacement.

2. With the hypothesis that coefficients m_{rs} of $W(Y)$ satisfy the conditions

$$m_{34} = m_{35} = 0 , \tag{7.117}$$

one may demonstrate that

$$
\begin{aligned}
Y'_{33} &= - \frac{1}{A^*} \left(\frac{R_1}{\varrho_1^2} y_1 + \frac{R_2}{\varrho_2^2} y_2 \right), \\
Y''_{33} &= - \frac{1}{A^*} \left[R_3 - \frac{M_2}{\varrho_1^2} y_1 + \frac{M_1}{\varrho_2^2} y_2 \right],
\end{aligned}
\right\} \tag{7.118}
$$

where

$$\varrho_i^2 = \frac{1}{A^*} \int_{A^*} y_i^2 \, dA^* \qquad (i = 1, 2) . \tag{7.119}$$

Let us consider the case of isotropic bodies or, if anisotropic, such that

$$m_{14} = m_{15} = m_{24} = m_{25} = m_{45} = m_{36} = m_{46} = m_{56} = 0 \tag{7.120}$$

(that is, rhombic, hexagonal, tetragonal, or monometric crystals). In this case it may be demonstrated that the Saint-Venant integrability conditions reduce to two:

$$
\begin{aligned}
[m_{44} Y_{23,1} - m_{55} Y_{13,2}]_{,1} &= 2 m_{13} \frac{R_2}{A^* \varrho_2^2}, \\
[m_{44} Y_{23,1} - m_{55} Y_{13,2}]_{,2} &= - 2 m_{23} \frac{R_1}{A^* \varrho_1^2},
\end{aligned}
\right\} \tag{7.121}
$$

and while Menabrea's theorem is not valid, the following minimum theorem holds (GRIOLI [7]): *the stress which is actually present in the body, according to Saint-Venant's theory, is characterized by* (7.106), (7.110), (7.118) *and by stress components* Y_{23}, Y_{13} *which minimize the functional*

$$B = \int_{A^*} \left\{ \frac{1}{m_{44}} \left[m_{44} Y_{23} - \frac{m_{13} R_2}{A^* \varrho_2^2} Q_3 \right]^2 + \frac{1}{m_{55}} \left[m_{55} Y_{13} - \frac{m_{23} R_1}{A^* \varrho_1^2} Q_4 \right]^2 \right\} dA^* \tag{7.122}$$

in the set [which I denote by I^*] *of the functions satisfying* (7.107), (7.116), (7.111), (7.115).

In (7.122) Q_3 denotes the sum of y_1^2 and an arbitrary polynomial of the first degree in y_1, y_2, while Q_4 is the sum of y_2^2 and of an arbitrary polynomial of second degree without the terms in y_2^2, $y_1 y_2$. The couple Y_{13}, Y_{23} which satisfies the above theorem also satisfies the integrability conditions (GRIOLI [7]).

2. Let $\{v_t\}$ be the sequence of all monomials formed by means of y_1, y_2 and ordered with non-decreasing degree, and let $\{P_t\}$ be such a sequence

of the independent polynomials obtained by adding to v_t that linear combination of $v_0, v_1, \cdots v_{t-1}$ which makes P_t orthogonal in $A*$ to such monomials. From (7.107), (7.116) the integral equations

$$\eta \overline{y_1^{\eta-1} y_2^{\tau} Y_{13}} + \tau \overline{y_1^{\eta} y_2^{\tau-1} Y_{23}} = c_{\eta\tau} \tag{7.123}$$

are obtained, where

$$c_{\eta\tau} = -\frac{1}{A*} \int\limits_{A*} \left[\frac{R_1}{\varrho_1^2} y_1 + \frac{R_2}{\varrho_2^2} y_2\right] y_1^{\eta} y_2^{\tau} dA* . \tag{7.124}$$

When $\eta = 0, \tau = 1$ or $\eta = 1, \tau = 0$, from (7.123) one deduces (7.111). Further, it follows that

$$\left. \begin{array}{ll} \overline{Y_{13} y_1} = \dfrac{c_{20}}{2} , & \overline{Y_{23} y_2} = \dfrac{c_{02}}{2} , \\[3mm] \overline{Y_{13} y_2} + \overline{Y_{23} y_1} = c_{11} . & \end{array} \right\} \tag{7.125}$$

which, together with (7.115), allows us to deduce the values of $\overline{Y_{13} y_2}$, $\overline{Y_{23} y_1}$.

Supposing $P_0 = 1, P_1 = y_1, P_2 = y_2$, let us put

$$\left. \begin{array}{l} Y'_{23} = -\dfrac{R_2}{A*} + \dfrac{1}{2\varrho_1^2}\left(c_{11} - \dfrac{M_3}{A*}\right) y_1 + \dfrac{c_{02}}{2\varrho_2^2} y_2 , \\[3mm] Y'_{13} = -\dfrac{R_1}{A*} + \dfrac{c_{20}}{2\varrho_1^2} y_1 + \dfrac{1}{2\varrho_2^2}\left(c_{11} + \dfrac{M_3}{A*}\right) y_2 , \end{array} \right\} \tag{7.126}$$

$$Y''_{r3m} = \sum_{t=3}^{m} \frac{\overline{Y_{r3} P_t}}{\varrho_t^2} P_t \qquad (r = 1, 2; m = 3, 4, \cdots), \tag{7.127}$$

$$\varrho_t^2 = \frac{1}{A*} \int\limits_{A*} P_t^2 dA* \qquad (t = 3, 4, \cdots) . \tag{7.128}$$

The special completeness property of the sequence $\{P_t\}$ assures that if $\overline{Y_r P_t}$ are constructed from linear combinations of solutions of (7.123) and are such that Y''_{r3m} is convergent for $m \to \infty$, then, putting

$$Y''_{r3} = \lim_{m \to \infty} Y''_{r3m} \qquad (r = 1, 2) , \tag{7.129}$$

the functions

$$Y_{r3} = Y'_{r3} + Y''_{r3} \qquad (r = 1, 2) , \tag{7.130}$$

satisfy equations (7.111), (7.115) and equations (7.107), (7.116) almost everywhere. It is possible to show that equations (7.123) for $\eta + \tau > 1$ are equivalent to the following:

$$\sum_{q=3}^{t'} \left[\alpha_{q1}^{(t)} \overline{Y_{13} P_q} + \alpha_{q2}^{(t)} \overline{Y_{23} P_q}\right] = \frac{1}{A*} \left[\alpha_{01}^{(t)} R_1 + \alpha_{02}^{(t)} R_2\right] - \frac{c_{20}}{2} \alpha_{11}^{(t)} -$$

$$- \frac{c_{02}}{2} \alpha_{12}^{(t)} - \frac{c_{11}}{2} \left(\alpha_{21}^{(t)} + \alpha_{12}^{(t)}\right) + \frac{M_3}{2A*} \left(\alpha_{12}^{(t)} - \alpha_{21}^{(t)}\right) \qquad (t=3,4,\cdots), \quad (7.131)$$

where

$$\alpha_{qr}^{(t)} = \frac{1}{A* \varrho_q^2} \int_{A*} P_q P_{t,r} \, dA* \qquad (q = 0, 1, \cdots; t = 3, 4, \cdots; r = 1, 2).$$
$$(7.132)$$

Putting

$$I'_m = A* \sum_{t=3}^{m} \frac{m_{44} \overline{Y_{23} P_t}^2 + m_{55} \overline{Y_{13} P_t}^2}{\varrho_t^2} -$$

$$- 2 \left[\frac{m_{13} R_2}{\varrho_2^2} \overline{Y_{23} Q_3} + \frac{m_{23} R_1}{\varrho_1^2} \overline{Y_{13} Q_4}\right], \qquad (7.133)$$

one may show that as a consequence of the previous minimum property the following theorem holds: *the solution of Saint-Venant's problem is given by the functions* (7.110), (7.118) *and* (7.126), (7.127), (7.129), (7.130) *if and only if* $\overline{Y_{r3} P_t}$ $(r = 1, 2)$ *minimize* I'_m *in the set of the solutions of equations* (7.131).

The above integration method is valid if the body's structure is such that (7.117), (7.120) hold. But one may demonstrate that if this hypothesis is discarded, Saint-Venant's problem is incompatible, except in some special cases. A. BRESSAN [2] has studied the question, determining the exceptional cases of compatibility[1].

For applications to the cases of simple extension, pure bending, torsion[2], and non-uniform bending, see GRIOLI [7].

---

[1] See also BRESSAN [1], where the possibility of expressing the solutions of Saint-Venant's problem for some types of anisotropic bodies by means of the corresponding ones of the isotropic case is examined. For comparison between isotropic and anisotropic cases see also LODGE [1, 2].

[2] For studying singularities which are present at a corner point of a prism with a hole, see PICONE [5] and FICHERA [6]. In particular, it is demonstrated that a stress concentration may be present at the corner $Q*$ if and only if the cross section is reentrant at $Q*$ and the region occupied by the body in the neighborhood of $Q*$ corresponds to an angle greater than π.

Chapter VIII

Plane Elasticity

§ 1. Preliminaries

1. It is not my aim to develop a general theory of plane elasticity. In this field, with reference to the linear case, there is a very large number of papers, while elasticity texts generally reserve many chapters for plane elasticity.

Naturally, the plane problems of elastic equilibrium may be studied by means of integral properties like those in Chapters VI and VII, but this is not my scope.

The questions concerning the existence and uniqueness theorems may be substantially expressed as analogous questions regarding harmonic and biharmonic functions, and there is the possibility of establishing theorems analogous to those of FICHERA [7] valid in the three-dimensional case.

With these theorems it is possible to construct polynomial solutions and to establish methods for the direct determination of the displacement components. Methods for studying the plane problems by means of complex variables are in MUSKHELISHVILI, while in SOBRERO [1] there is a new algorithm for studying the plane problem, based on the use of hyper-complex numbers. A book concerning plane linear systems in equilibrium or steady motion has just been published by MILNE-THOMSON.

Supposing that body forces are absent, as it is generally possible to assume, it is well known that the linear plane problems of elastic equilibrium may be reduced to the determination of a biharmonic function: *Airy's stress function*.

In this chapter I wish only to state some considerations regarding its analytical structure.

First I wish to call attention to the fact that some cases exist for which the stress of the plane problem depends only on a harmonic function—and this case places stringent conditions on the external forces—or on a harmonic function and a biharmonic one.

It is appropriate to find the analytical structure of Airy's function in order to study the integration problem even in the case of a multiply connected region; in such manner its determination may be reduced to the determination of regular functions.

2. Two quite different cases occur, even if they often derive from the same analytical problem.

Case I: plane deformation. The displacements are parallel to the $y_1 \, y_2$ plane, $u_3 \equiv 0$. As is well known, the u_1, u_2 do not depend on y_3, and the stress depends on a biharmonic function.

Specifically

$$Y_{11} = H_{,22}, \qquad Y_{22} = H_{,12}, \qquad Y_{12} = -H_{,12}, \qquad (8.1)$$

$$Y_{33} = \Delta_2 H, \qquad Y_{13} = 0, \qquad Y_{23} = 0, \qquad (8.2)$$

with $H(y_1, y_2)$ satisfying the equation

$$\Delta_2 \Delta_2 H = 0 \qquad \left[\Delta_2 = \frac{\partial^2}{\partial y_1^2} + \frac{\partial^2}{\partial y_2^2} \right]. \qquad (8.3)$$

Case II: plane stress. The stress satisfies the conditions

$$Y_{13} = Y_{23} = Y_{33} = 0. \qquad (8.4)$$

If no other condition is added to (8.4), the stress, as is well known, depends on y_3 and is in general expressible by a biharmonic function and a harmonic one.

Specifically, if E denotes Young's modulus and ν Poisson's coefficient, (8.1) is valid but H is expressed by

$$H(y_1, y_2, y_3) = B_1(y_1, y_2) + B_2(y_1, y_2) y_3 - \frac{\nu y_3^2}{2(1+\nu)} \Delta_2 B_1(y_1, y_2), \quad (8.5)$$

where $B_1(y_1, y_2)$ is a biharmonic function and $B_2(y_1, y_2)$ a harmonic one. The case of plane stress requires particular attention if the stress must not depend on y_3, as is often required in the case of the equilibrium of cylindrical bodies when the surface forces acting only upon the lateral surface are independent of y_3 and are perpendicular to the axis $O^* y_3$. In this case (8.1) is still valid, but H cannot be expressed by (8.5), except in the special case $\nu = 0$, for which $B_2 \equiv 0$ is to be assumed. That is, H cannot be a biharmonic function, although, strangely, the contrary is stated in some theoretical elasticity books.

The problem of plane stress with Y_{rs} independent of y_3 has been studied by A. GHIZZETTI, who, using the displacement components, has shown that the stress depends only on a harmonic function[1]. This fact, valid if $\nu \neq 0$, is strongly restrictive for the surface forces, and the problem is compatible only in special cases. Airy's function H is biharmonic only if some of the Saint-Venant integrability conditions are neglected.

The problem of plane stress independent of y_3 is readily treated by using the Saint-Venant integrability conditions expressed by means of the stress components, but it is substantially contained as a particular case in the general case of plane stress.

---

[1] However, the stress corresponding to a biharmonic function H may be a good approximation in the case of thin plates [TIMOSHENKO and GOODIER, p. 241].

In fact, if one requires Y_{11}, Y_{22}, Y_{12} to be independent of y_3, from (8.1), (8.2), (8.5) follows

$$B_2(y_1, y_2) \equiv 0, \qquad \nu \Delta_2 B_1(y_1, y_2) \equiv 0. \qquad (8.6)$$

That is, Airy's function is biharmonic, as in the case of plane deformations, only if $\nu = 0$; if $\nu \neq 0$, $H(y_1, y_2)$ is expressed by[1]

$$H(y_1, y_2) = B_1(y_1, y_2) + \frac{1}{6}(\varrho_1 y_1^3 + \varrho_2 y_2^3) + \frac{1}{4}\varrho_3(y_1^2 + y_2^2), \qquad (8.7)$$

with $B_1(y_1, y_2)$ harmonic, in accordance with GHIZZETTI's result[2].

§ 2. Structure of Airy's stress function in the problem of plane deformations

1. It is interesting to put in evidence the singularities of AIRY's function in order to reduce the analytical problem to one of determining functions which are regular in the plane region where the body is defined. It is possible to show that the coefficients of the singular terms have a precise physical meaning or have some properties which express known average properties of stress. For details see GRIOLI [2].

To simplify, I shall consider the case of a doubly connected region A^*, but the results may be extended to general multiply connected regions (GRIOLI [2]).

Let l^* be any closed curve contained in A^* which goes round the hole whose contour is σ_1^*. Further, let R_1, R_2 be the components with respect to axes y_1, y_2 of the resultant of the forces acting upon σ_1^* and M their resultant moment with respect to the axis of y_3.

From (8.1) follows[3]

$$\int_{l^*} d(H_{,2}) = -R_1, \qquad \int_{l^*} d(H_{,1}) = R_2, \qquad (8.8)$$

$$\int_{l^*} d[H - y_1 H_{,1} - y_2 H_{,2}] = -M. \qquad (8.9)$$

Equations (8.8), (8.9) show that the expressions contained in the integrals are, in general, multi-valued functions unless the external forces take a particular form.

---

[1] The terms in ϱ_1, ϱ_2, ϱ_3 are not written explicitly in (8.5), since they may be included in the biharmonic function $B(y_1, y_2)$.

[2] For an application of such results to the case of a cylindrical body with specified normal forces acting upon the lateral surface, see PLATONE.

[3] Formulae (8.8) are in accordance with a well known property of Airy's function [SOBRERO, 3, 4]. See also TOLOTTI [1].

Generally, $H(y_1, y_2)$ is a multi-valued function together with its first derivatives, but its second derivatives are single-valued, since they represent the stress components.

The function $\theta = \operatorname{arctg} \dfrac{y_2}{y_1}$ is harmonic with single-valued derivatives, and it is easy to recognize that the function

$$\Psi(y_1, y_2) = -\frac{M'}{2\pi}\,\theta\,, \tag{8.10}$$

where

$$M' = M + y_2 R_1 - y_1 R_2\,, \tag{8.11}$$

satisfies (8.8), (8.9).

Clearly M' is the resultant moment with respect to the point $P^* \equiv (y_1, y_2)$ of the external forces acting upon σ_1^*. It is possible to show that the form of H is necessarily

$$H(y_1, y_2) = \varphi(y_1, y_2) - \frac{M'}{2\pi}\,\theta\,, \tag{8.12}$$

where $\varphi(y_1, y_2)$ is a biharmonic and single-valued function in A^*.

2. It is convenient to use for $\varphi(y_1, y_2)$ the decomposition formula in the POINCARÈ manner (FICHERA [1]):

$$\varphi(y_1, y_2) = \varphi_0(y_1, y_2) + \varphi_1(y_1, y_2) + [\alpha \varrho^2 + \beta y_1 + \gamma y_2 + \delta]\log \varrho +$$
$$+ a \cos 2\theta + b \sin 2\theta\,, \tag{8.13}$$

where $\varrho = \sqrt{y_1^2 + y_2^2}$. The function φ_0 is biharmonic in the domain bounded by the exterior contour σ_0^* of A^*, and φ_1 is biharmonic at the exterior of the domain bounded by σ_1^* and convergent at infinity[1]. To specify the constants α, β etc., it is convenient to consider the operator

$$\Phi[u, v, l^*] = \frac{1}{8\pi}\left\{\int_{l^*}\left[v\,\frac{d\Delta_2 u}{dn} - u\,\frac{d\Delta_2 v}{dn}\right]ds + \int_{l^*}\left[\Delta_2 v\,\frac{du}{dn} - \Delta_2 u\,\frac{dv}{dn}\right]ds\right\}. \tag{8.14}$$

---

[1] By known theorems, the functions φ_0, φ_1 may be expanded in power series,

$$\varphi_0 = \sum_{m=0}^{\infty} \varrho^m (T_{12} + \varrho^2 T_{2n}), \qquad \varphi_1 = \sum_{m=3}^{\infty} \frac{1}{\varrho^n}(T_{3n} + \varrho^2 T_{4n}) + \frac{T_{31}}{\varrho} + \frac{T_{32}}{\varrho^2}\,, \tag{*}$$

with

$$T_{in} = a_{in}\cos n\theta + b_{in}\sin n\theta \qquad (i = 1, 2, 3, 4;\quad n = 0, 1, \ldots).$$

In this case the expression (8.13) for φ coincides with the solution given by J. H. MICHELL and is particularly useful in finding the stress in a circular ring. See also TIMPE. The series (*) are uniformly convergent, the first one in every finite region, the second for $\varrho > 0$. In the second of (*) the term in T_{42} is absent, as is explicitly written in (8.13).

If u, v are biharmonic functions in A^*, the value of $\Phi[u, v, l^*]$ does not depend on the curve l^*, and there results

$$\alpha = \Phi[\varphi, 1, l^*], \quad \beta = -2\Phi[\varphi, y_1, l^*], \quad \gamma = -2\Phi[\varphi, y_2, l^*],$$
$$\delta = \Phi[\varphi, \varrho^2, l^*], \quad a = \frac{1}{2}\Phi[\varphi, \varrho^2\cos 2\theta, l^*], \quad b = \frac{1}{2}\Phi[\varphi, \varrho^2\sin 2\theta, l^*]. \quad \left.\right\} \quad (8\;15)$$

§ 3. Meaning of the coefficients of the singular part of Airy's function in the case of plane deformations

1. Keeping in mind the known formula which in the linear case relates the stress components to the strain components, by easy calculations one deduces[1]

$$u_1(y_1 y_2) = u_1(y_1^{(0)}, y_2^{(0)}) - [y_2\,\omega\,(y_1, y_2) - y_2^{(0)}\,\omega\,(y_1^{(0)}, y_2^{(0)})] + $$
$$+ \frac{1}{E}\left\{ \tau\int_{l_1^*}\left[y_2\frac{d\Delta_2 H}{dn} - \Delta_2 H\frac{dy_2}{dn}\right]ds - (1+\nu)\int_{l_1^*} d(H_{,1}) \right\},$$
$$u_2(y_1, y_2) = u_2(y_1^{(0)}, y_2^{(0)}) + [y_1\,\omega\,(y_1, y_2) - y_1^{(0)}\,\omega\,(y_1^{(0)}, y_2^{(0)})] - $$
$$- \frac{1}{E}\left\{ \tau\int_{l_1^*}\left[y_1\frac{d\Delta_2 H}{dn} - \Delta_2 H\frac{dy_1}{dn}\right]ds + (1+\nu)\int_{l_1^*} d(H_{,2}) \right\}, \quad \left.\right\} \quad (8.16)$$
$$\omega\,(y_1, y_2) = \omega\,(y_1^{(0)}, y_2^{(0)}) + \frac{\tau}{E}\int_{l_1^*}\frac{d\Delta_2 H}{dn}ds,$$

where $\omega\,(y_1, y_2)$ is the local rotation, the path of integration is, a line l_1^* contained in A^*, starting from $P_0 \equiv (y_1^{(0)}, y_2^{(0)})$ and ending at $P \equiv (y_1, y_2)$, and

$$\tau = 1 - \nu^2. \quad (8.17)$$

2. Let the body be subjected to a plane dislocation, that is, to a dislocation which affects only the displacement components u_1, u_2. It may be made by cutting the body in such manner that the region A^* becomes simply connected and then displacing one of the cut's edges relative to the other one by a translation parallel to the plane $O^* y_1 y_2$ and by a rotation around an axis perpendicular to this plane, possibly adding or taking away a certain quantity of material so as to make the two edges meet. Therefore, the characteristic parameters of a plane dislocation are three. I shall prove this theorem: *In the expression of Airy's function determined by (8.11), (8.12), (8.13), the constants α, β, γ characterize the three characteristic parameters of the plane dislocation and are in turn specified by it*[2].

---

[1] In regard to the possible expansion of u_i, ω in series of the type (*) in footnote 1, p. 130, see GRIOLI [2].

[2] For results regarding plane dislocations of a region bounded by two concentric circles, see TIMPE.

Let us begin by observing that from (8.16) it follows that necessary and sufficient conditions for the single-valuedness of the displacement components are

$$\int_{i^*} \frac{d\Delta_2 H}{dn} \, ds = 0 \, ,$$

$$\tau \int_{i^*} \left[y_2 \frac{d\Delta_2 H}{dn} - \Delta_2 H \frac{dy_2}{dn} \right] ds - (\nu + 1) \int_{i^*} d(H_{,1}) = 0 \, ,$$

$$\tau \int_{i^*} \left[y_1 \frac{d\Delta_2 H}{dn} - \Delta_2 H \frac{dy_1}{dn} \right] ds + (\nu + 1) \int_{i^*} d(H_{,2}) = 0 \, .$$

$\qquad\qquad$ (8.18)

The functions $\varphi_0(y_1, y_2)$, $\varphi_1(y_1, y_2)$ satisfy (8.18). This is evident for (8.18, 1), since φ_0, φ_1 are harmonic functions. With reference to (8.18, 2), from

$$\left(y_2 \frac{d\Delta_2 H}{dn} - \Delta_2 H \frac{dy_2}{dn} \right) ds = y_2 (\Delta_2 H)_{,1} \, dy_2 - [y_2(\Delta_2 H)_{,2} - \Delta_2 H] \, dy_1$$

$\qquad\qquad$ (8.19)

it follows that the first member of (8.19) is an exact differential, since H is a biharmonic function. Then the left-hand member of (8.18, 2) is the sum of integrals of exact differentials, and, since the field of regularity of φ_0 is the domain bounded by σ_0^*, one shows that it is equal to zero for $H \equiv \varphi_0$. The same thing is shown for $H \equiv \varphi_1$ if one considers that its field of regularity is formed by all points not interior to the domain bounded by σ_1^*. Analogously for (8.18, 3).

The functions $\log \varrho$ and θ satisfy (8.18) since they are harmonic, the first being single-valued together with its first derivatives, the second being multi-valued but with single-valued first derivatives. Also the functions $\cos 2\theta$, $\sin 2\theta$ satisfy (8.18), as it is easy to show.

Instead, the functions $\varrho^2 \log \varrho$, $y_1 \log \varrho$, $y_2 \log \varrho$, $y_1 \theta_1$, $y_2 \theta$ do not satisfy the single-valuedness conditions (8.18). Consequently, of the terms contained in Airy's function [see (8.11), (8.12), (8.13)], only those belonging to the expression

$$H^*(y_1, y_2) = (\alpha \varrho^2 + \beta y_1 + \gamma y_2) \log \varrho + \frac{1}{2\pi} (y_1 R_2 - y_2 R_1) \theta \quad (8.20)$$

give multi-valued displacements.

Denoting by u_1', u_2', ω' the increases in u_1, u_2, ω when one makes a complete circuit round the hole starting from the point with coordinates y_1, y_2, from (8.16) it follows that $E\omega'$, $E(u_1' + y_2\omega')$, $E(u_2' - y_1\omega')$ coincide with the values which the first members of (8.18) assume when

$H \equiv H^*$. Therefore we have

$$
\left.
\begin{aligned}
u_1' &= -2\pi \frac{\tau}{E} \left[4\alpha y_2 + 2\gamma - \frac{R_2}{\pi}\left(1 - \frac{1+\nu}{2\tau}\right) \right], \\
u_2' &= 2\pi \frac{\tau}{E} \left[4\alpha\gamma_1 + 2\beta - \frac{R_1}{\pi}\left(1 - \frac{1+\nu}{2\tau}\right) \right], \\
\omega' &= 8\pi \frac{\tau}{E}\alpha .
\end{aligned}
\right\}
\qquad (8.21)
$$

Since the external forces vanish for any dislocation, we must put $R_1 = R_2 = M = 0$.

Then from (8.21) follows: *The three characteristic parameters of the plane dislocation are given by*

$$
-4\pi \frac{\tau}{E}\gamma , \qquad 4\pi \frac{\tau}{E}\beta , \qquad 8\pi \frac{\tau}{E}\alpha . \qquad (8.22)
$$

The first two correspond to uniform fissures, the third to a radial[1] one, and they are to be thought of as specified, corresponding to the considered dislocation. Then if l, m, r denote the specified characteristic parameters of the dislocation, one has: *Airy's stress function corresponding to a plane dislocation of the characteristic parameters l, m, r in a doubly connected region may be expressed in the form*

$$
H = \varphi_0 + \varphi_1 + \frac{E}{4\pi\tau}\left[-\frac{r}{2}\varrho^2 + my_1 - ly_2 \right]\log \varrho + \delta \log \varrho +
$$
$$
+ a\cos 2\theta + b\sin 2\theta . \qquad (8.23)
$$

The unknown part $H' = H - H^*$ of the second member of (8.23) is restricted only by the condition that the boundary forces all vanish. Explicitly, such a condition is

$$
\left.
\begin{aligned}
H'_{,1} &= -H^*_{,1} + \text{const.} \\
H'_{,2} &= -H^*_{,2} + \text{const.}
\end{aligned}
\right\}
\quad (\text{on } \sigma_0^* \text{ and } \sigma_1^*) . \qquad (8.24)
$$

3. Now, let us suppose that dislocations are absent. It is easy to show that coefficients α, β, γ are still determined. In fact, the single-valuedness conditions, which are

$$
u_1' = 0 , \qquad u_2' = 0 , \qquad \omega' = 0 , \qquad (8.25)
$$

---

[1] Clearly, such definitions represent a generalisation of those used in V. VOLTERRA's papers in the case of a region bounded by two concentric circles. That is, the dislocation corresponds to a uniform fissure if the corresponding displacement is a translation, a radial fissure if it is a rotation.

according to (8.21) give

$$\alpha = 0. \qquad \beta = \frac{R_1}{2\pi}\left(1 - \frac{1+\nu}{2\tau}\right), \qquad \gamma = \frac{R_2}{2\pi}\left(1 - \frac{1+\nu}{2\tau}\right). \quad (8.26)$$

From (8.11), (8.12), (8.13), (8.26) follows: *In the case of single-valued displacements and in a doubly connected region Airy's function may be expressed by*

$$H = \varphi_0 + \varphi_1 + \delta \log \varrho + a \cos 2\theta + b \sin 2\theta + \frac{1}{2\pi}\Big[(R_1 y_1 +$$

$$+ R_2 y_2)\left(1 - \frac{1+\nu}{2\tau}\right) \log \varrho - M'\theta\Big]. \quad (8.27)$$

From (8.27) it follows that a necessary and sufficient condition for Airy's stress function in a doubly connected region to be independent of the elastic coefficients is that the external forces acting upon the contour of the hole have zero resultant. The same holds for the stress components.

4. The coefficients δ, a, b are not determined, as are α, β, γ, by the fundamental elements of the problem. However, they have a special physical meaning and may be expressed precisely by some characteristic elements of deformation and stress, such as the local rotation, the surface dilatation coefficient and the astatic coordinates of external forces acting upon one of the two contours of the plane region. Specifically, it may be demonstrated that if δ_A is the surface dilatation coefficient, then

$$\delta = \frac{E}{8\pi\left[\tau\,(\tau+\nu^2)-\nu\,(1+\nu)\right]}\left[\int_{l^*} \varrho^2 \frac{d\,\delta_A}{dn}\,ds -\right.$$

$$\left. -\int_{l^*} \delta \frac{d\,\varrho^2}{dn}\,ds\right] - \frac{A_{l^*}}{2\pi}\,(a_{11}+a_{22}), \quad (8.28)$$

where a_{11}, a_{22} are two astatic coordinates of the forces acting through l^* and A_{l^*} the area enclosed by l^*. Further, one finds

$$a = \frac{E}{16\pi\left[\tau\,(\tau+\nu^2)-\nu\,(1+\nu)\right]}\left\{\int_{l^*}(y_1^2-y_2^2)\frac{d\,\delta_A}{dn}\,ds -\int_{l^*}\frac{d\,(y_1^2-y_2^2)}{dn}\,\delta_A\,ds\right\},$$

$$b = \frac{E}{8\pi\left[\tau\,(\tau+\nu^2)-\nu\,(1+\nu)\right]}\left\{\int_{l^*}y_1\,y_2\frac{d\,\delta_A}{dn}\,ds -\int_{l^*}\frac{d\,y_1\,y_2}{dn}\,\delta_A\,ds\right\}.$$

$$\text{(8.29)}$$

If one of the two contours of the plane region is free of external forces, it is sufficient to suppose that l^* coincides with that contour in order that the term with $a_{11}+a_{22}$ be absent in the expression of δ. If one wishes, it is possible to express δ, a, b by means of the local rotation.

Equations (8.28), (8.29) may be interpreted in several ways. In particular, their invariance expresses known average properties of stress, analogous to those contained in Chapter VI.

§ 4. Structure of Airy's function in the problem of plane stress independent of y_3

1. As has been observed, if $v = 0$ the stress is represented by means of a biharmonic function, as in the case of plane deformations, and this is natural, since if $v = 0$, the general formulae are the same in both cases.

Instead, if $v \neq 0$, Airy's function is expressed by (8.7) with B_1 harmonic. (8.16) are still valid provided one puts 1 in the place of τ. B_1 being harmonic, from (8.7) follows

$$
\begin{aligned}
u_1 =& -\frac{1+v}{E} \int_{i^*} d(B_{1,1}) - \frac{1}{E}\left\{\frac{1}{2}\varrho_1[v(y_1^2 - y_3^2) + y_2^2] - \right. \\
& \left. - \varrho_2 y_1 y_2 - \frac{1-v}{2}\varrho_3 y_1 \right\}, \\
u_2 =& -\frac{1+v}{E} \int_{i^*} d(B_{1,2}) - \frac{1}{E}\left\{\frac{1}{2}\varrho_2[v(y_2^2 - y_3^2) + y_1^2] - \right. \quad (8.30) \\
& \left. - \varrho_1 y_1 y_2 - \frac{1-v}{2}\varrho_3 y_2 \right\}, \\
u_3 =& -\frac{v}{E}(\varrho_1 y_1 + \varrho_2 y_2 + \varrho_3)\, y_3,
\end{aligned}
$$

$$
\omega_1 = -\frac{v}{E}\varrho_2 y_3, \qquad \omega_2 = \frac{v}{E}\varrho_1 y_3, \qquad \omega_3 = \frac{1}{E}(\varrho_1 y_2 - \varrho_2 y_1), \quad (8.31)
$$

where ω_i are the components of the local rotation.

From (8.8), (8.9), (8.30) one deduces that the increase u_i', ω_i' in u_i, ω_i when one makes a circuit around the hole are

$$
u_1' = -\frac{1+v}{E} R_2, \qquad u_2' = \frac{1+v}{E} R_1, \qquad u_3' = 0, \qquad \omega_i' = 0. \quad (8.32)
$$

(8.32) show that the existence of plane dislocations is impossible with plane stress independent of y_3, since $u_1' = u_2' = 0$ when $R_1 = R_2 = 0$. Further, it is evident that the displacement u cannot be a single-valued function if R_1, R_2 are not simultaneously zero. Therefore: *A necessary and sufficient condition for an equilibrium state with plane stress independent of y_3 is that the forces acting upon each of the contours have their resultants separately equal to zero.*

Consequently, $H(y_1, y_2)$ has single-valued first derivatives. Keeping in mind (8.11), one deduces that B_1 is the sum of multi-valued functions $-\dfrac{M}{2\pi}\theta$ and a single-valued harmonic function in A^*. Then using for the latter a well known decomposition formula (PICONE [1]), there results

$$
H(y_1, y_2) = -\frac{M}{2\pi}\theta + \varphi(y_1, y_2) + \frac{1}{6}(\varrho_1 y_1^3 + \varrho_2 y_2^3) + \frac{1}{4}\varrho_3(y_1^2 + y_2^2),
$$
$$
(8.33)
$$

with

$$\varphi(y_1, y_2) = \bar{\delta} \log \varrho + \Psi_0(y_1, y_2) + \Psi_1(y_1, y_2) \,. \qquad (8.34)$$

In (8.34) $\bar{\delta}$ is a constant, Ψ_0 a function harmonic within the domain bounded by σ_0^*, and Ψ_1 a function harmonic at all points external to the domain bounded by σ_1^* and convergent at infinity.

2. The meaning of $\bar{\delta}$ is still given by (8.28), where, however, δ_A depends only on ϱ_3, as is easily recognized. It is evident that the external forces cannot be specified arbitrarily, even if they have a null resultant upon each contour. From (8.33) one recognizes that the set of external forces may be thought of as composed of the superposition of two sets, of which the first is determined by the vector f' with components

$$f_1' = \left(\varrho_2 y_2 + \frac{1}{2}\,\varrho_3\right) N_1^*\,, \qquad f_2' = \left(\varrho_1 y_1 + \frac{1}{2}\,\varrho_3\right) N_2^* \qquad \text{(on } \sigma_0^* \text{ and } \sigma_1^*) \qquad (8.35)$$

and the second by the vector f'' whose components are

$$\begin{aligned}
f_1'' &= \left(\varphi - \frac{M}{2\pi}\,\theta\right)_{,22} N_1^* - \left(\varphi - \frac{M}{2\pi}\,\theta\right)_{,12} N_2^* \\
f_2'' &= -\left(\varphi - \frac{M}{2\pi}\,\theta\right)_{,12} N_1^* + \left(\varphi - \frac{M}{2\pi}\,\theta\right)_{,11} N_2^* \,.
\end{aligned} \quad \left.\begin{array}{r} \text{(on } \sigma_0^* \text{ and } \sigma_1^*)\,, \end{array}\right\} \quad (8.36)$$

For example, neglecting unimportant constants, from (8.36) we see that

$$\frac{d\varphi}{dN^*} = F(s) \qquad \text{(on } \sigma_0^* \text{ and } \sigma_1^*)\,, \qquad (8.37)$$

where

$$F(s) = \frac{M}{2\pi}\,\frac{d\theta}{dN^*} - N_1^* \int_0^s f_2''(\zeta)\,d\zeta + N_2^* \int_0^s f_1''(\zeta)\,d\zeta\,. \qquad (8.38)$$

Therefore φ is determined to within inessential constants by a Neumann problem after specification of the vector $N^* \times \int_0^s f''(\zeta)\,d\zeta$ on the boundary of the plane region.

§ 5. Structure of Airy's function in the problem of plane stress depending on y_3

The case of plane stress depending on y_3 may be studied by a criterion of superposition, keeping in mind (8.5). It is sufficient to consider the body section $y_3 = 0$ to deduce that the biharmonic function $B_1(y_1, y_2)$ has the same analytical structure as $H(y_1, y_2)$ in the case of plane deformations, provided $\tau = 1$. Let us denote by $R_i(y_3)$ $(i = 1, 2)$ the components of the resultant of the external forces acting upon the interior contour of the section $y_3 =$ const., and by $M\ (y_3)$ the components of their resultant moment with respect to the y_3 axis. Therefore, considering for simplicity the case of a single-valued displacement, we have

$$B_1\,(y_1, y_2) = \varphi_0 + \varphi_1 + \delta \log \varrho + a \cos 2\theta + b \sin 2\theta + \frac{1}{2\pi} \Big\{ [y_1 R_1(0) +$$

$$+ y_2 R_2(0)]\,\frac{1-\nu}{2} \log \varrho - [M\,(0) + y_2 R_1(0) - y_1 R_2(0)]\,\theta \Big\}. \qquad (8.39)$$

Considerations analogous to those made about $H(y_1, y_2)$ in the case of plane stress independent of y_3 may be repeated for $B_2(y_1, y_2)$. Thus one deduces that the external forces corresponding to $B_2(y_1, y_2)$ have resultant equal to zero on $\sigma_1^*\,(y_3)$. It follows that

$$B_2(y_1, y_2) = -\frac{p}{2\pi}\,\theta + \Psi_0 + \Psi_1 + \delta \log \varrho + \frac{1}{6}\,(\varrho_1 y_1^3 + \varrho_2 y_2^3) +$$

$$+ \frac{1}{2}\,\varrho_3\,(y_1^2 + y_2^2)\,, \qquad (8.40)$$

where p is a constant.

The function $\varDelta_2 B_1$, being single-valued, corresponds to forces with resultant and resultant moment equal to zero, acting through any closed line l^* going around the hole. Consequently, the external forces must have the same resultant, independent of y_3, on each contour of the section $y_3 =$ const. If $\bar{u}_1, \bar{u}_2$ denote the displacement components due only to the term B_1, it is easily deduced that

$$u_1 = \bar{u}_1 - \frac{1+\nu}{E}\,B_{2,1}y_3 + \frac{\nu}{2E}\,(\varDelta_2 B_1)_{,1}y_3^2\,,$$

$$u_2 = \bar{u}_2 - \frac{1+\nu}{E}\,B_{2,2}y_3 + \frac{\nu}{2E}\,(\varDelta_2 B_1)_{,2}y_3^2\,, \qquad (8.41)$$

$$u_3 = \frac{1+\nu}{E}\,B_2 - \frac{\nu}{E}\,\varDelta_2 B_1 y_3\,.$$

The single-valuedness of u_1, u_2 is assured. Instead, the single-valuedness of u_3 implies the single-valuedness of B_2, and therefore [see (8.40)] $p = 0$. Then the external forces must have the same resultant moment with respect to the y_3 axis, independent of y_3, on each contour of the section $y_3 =$ const. It is easy to show that the single-valuedness of the local rotation follows. Therefore, in the case of single-valued displacements, the expression of the Airy function is

$$H\,(y_1, y_2, y_3) = \varphi_0 + \varphi_1 + \delta \log \varrho + a \cos 2\theta + b \sin 2\theta +$$

$$+ \frac{1}{2\pi}\Big\{ [y_1 R_1 + y_2 R_2]\,\frac{1-\nu}{2} \log \varrho - [M + y_2 R_1 - y_1 R_2]\,\theta \Big\} +$$

$$+ \Big[\Psi_0 + \Psi_1 + \delta \log \varrho + \frac{1}{6}\,(\varrho_1 y_1^3 + \varrho_2 y_2^3) + \frac{1}{4}\,\varrho_3\,(y_1^2 + y_2^2) \Big]\,y_3 -$$

$$- \frac{\nu}{2\,(1+\nu)}\,\Big[\varDelta_2 \varphi_0 + \varDelta_2 \varphi_1 - \frac{4}{\varrho^2}\,(a \cos 2\theta + b \sin 2\theta) -$$

$$- \frac{1+\nu}{2\pi\varrho^2}\,(R_1 y_1 + R_2 y_2) \Big]\,y_3^2\,. \qquad (8.42)$$

It is evident how restrictive the case is: the external forces depend on second-degree polynomials in y_3 and on each contour have resultant and resultant moment independent of y_3 and independent of the contour. Further, in these polynomials the coefficient of the term in y_3^2 is determined by the external forces acting upon the section $y_3 = 0$, while that of the term in y_3 depends on a harmonic function and therefore is subjected to very restrictive conditions, like those expressed in the previous paragraph.

Since the plane stress problem is so restrictive, it is often changed into the well known *generalized plane stress problem* [LOVE, p. 95], acceptable for cylindrical bodies of small height. In this case, the problem becomes analogous to the one of plane strain, and the solution depends on a biharmonic function, for which everything said in §§ 2, 3 is valid.

Chapter IX

Hypo-Elasticity

§ 1. Preliminaries

The most common way of treating the mechanics of continuous bodies and, in particular, the mechanics of elastic bodies, is based on the assumption of the existence of a free equilibrium state, which generally is an unstressed state. That is, one postulates the existence of an equilibrium state without external forces and, generally, without stress. The deformations are measured from this state C^*. In effect, assuming the existence of a natural state of equilibrium [without stress] determines a first property of the thermodynamic potential: Y_{rs} derived from it must be zero at C^*. At any rate, dropping the condition that C^* be unstressed, it is recognized that the stress is necessarily a uniform pressure if the body is homogeneous and isotropic in C^*, while it is a uniform stress if the body is anisotropic. It may sometimes be convenient to measure the deformations from a state for which the stress is known but depends on the coordinates in a complicated manner, as happens, for example, in the case of VOLTERRA's dislocations. It is often convenient to assume such a stressed reference state and to measure deformations from it. The common way of studying elastic problems is acceptable and convenient when it is easy and useful to assume an unstressed state as reference.

Since it is often not so, several authors (see, for example, ZAREMBA, JAUMANN, HANDELMAN, LIN and PRAGER, HENCKY [2]) suggest dropping the idea that an unstressed state necessarily exists and substituting for the usual stress-strain relations other relations between the rate of stress and the rate of strain, corresponding to a transformation from C to an infinitely close configuration.

From this procedure results a substantially dynamic theory in which stress is determined not only by the comparison between C and a reference configuration but also by the knowledge of intermediate states which the body goes through, starting from an arbitrary initial configuration. Now the constitutive equations represent relations between X_{rs}, the time derivatives of X_{rs}, and the velocities. For example, according to MUR-NAGHAN's [3, 4] theory in the case of isotropic bodies, one has

$$\dot{X}_{rs} = \lambda \delta_{rs} \dot{d}_p^{\,p} + 2\mu d_{rs},\tag{9.1}$$

where the dot denotes differentiation with respect to time and[1]

$$d_{rs} = \frac{1}{2}\left(v_{r/s} + v_{s/r}\right).\tag{9.2}$$

In (9.2) v_r are the components of the Eulerian velocity.

By substituting $\dot{X}_{rs}\,dt$ and $d_{rs}\,dt$ for X_{rs}, ε_{rs}, it is evident that (9.1) is equivalent to Hooke's law. Also, it is clear that an arbitrary initial stress may be assumed to be present; that is, the stress is arbitrary in the initial configuration, which now takes the place of the reference configuration. A theory based on the rate of stress and corresponding to the elasticity of large deformations is the theory of *hypo-elasticity*, proposed and developed by C. TRUESDELL [4, 5, 6, 7].

This theory is the most complete one based on the rate of stress and has found interest at once among several authors. I shall briefly consider it.

Hypo-elasticity is equivalent to the classical theory of elasticity only in the case of small deformations, while it may generally be regarded as an enlargement of the common elasticity theory of large strain to which hypo-elasticity is equivalent only in special cases. An interesting result of hypo-elasticity is that in some cases plastic yield may be considered the consequence of the basic equations. In other words, according to C. TRUES-DELL's opinion, hypo-elasticity *must* foresee plastic yield without the necessity of a separate *plasticity condition* which divides the elastic state from the plastic one.

§ 2. Basic equations of hypo-elasticity

One substitutes for the Eulerian components of stress X_{rs} the dimensionless quantity

$$\xi_{rs} = \frac{X_{rs}}{2\mu},\tag{9.3}$$

---

[1] In (9.2) and in the following equations an oblique dash (/) before an index r denotes differentiation with respect to x_r.

where μ is the one of the two Lamé coefficients of classic elasticity which cannot be zero. The constitutive equations of hypo-elasticity are of the form

$$\frac{\delta \xi^{rs}}{\delta t} = \varphi^{rs}(d_{pq}, \xi^{lm}) . \tag{9.4}$$

According to TRUESDELL's idea, no one of the moduli characterizing the medium may have dimensions which are independent of those of stress, and, further, the φ^{rs} are analytic functions of d_{pq} and ξ^{lm} whose coefficients are dimensionless. With such restrictions and by dimensional analysis one deduces that the φ^{rs} are linear homogeneous functions of d_{pq}, and (9.4) becomes

$$\frac{\delta \xi^{rs}}{\delta t} = A^{rspq} d_{pq} , \tag{9.5}$$

where the A^{rspq} are functions of ξ^{lm} only, whose coefficients are dimensionless.

2. It is very important to clarify the meaning of $\delta \xi^{rs}$. Certainly, $\delta \xi^{rs}$ cannot be such that the left-hand members of (9.5) are identified with the derivatives with respect to the time:

$$\dot{\xi}^{rs} = \frac{\partial \xi^{rs}}{\partial t} + \xi^{rs}_{/q} v^q . \tag{9.6}$$

In fact, if it were so, in the case of a rigid motion of the body, X_{rs} referred to a coordinate system fixed in the body would depend on the time, and this is impossible. One may determine the correct expressions of the left-hand members of (9.5) following W. NOLL's theory, who, studying in general the mechanics of continuous media, has established a general principle of the *isotropy of physical space*.

But it may be observed that the property that $\dfrac{\delta \xi^{rs}}{\delta t}$ be zero for any rigid motion[1] is equivalent to the vanishing of the time derivatives of the Lagrangean components of stress with reference to a configuration $\bar{C}$ chosen among those possible for the body. One considers a displacement from $\bar{C}$ depending on the parameter θ in such a way that to $\theta = 0$ corresponds the configuration $\bar{C}$. If C^* of Chapter IV is identified with $\bar{C}$, the relations (4.49) are valid. If $\bar{C}$ is the initial position of the body and θ is identified with the time t, (4.49, 1) becomes

$$[\dot{X}_{rs}]_{(0)} = [\dot{Y}_{rs}]_{(0)} + v^{(0)}_{r,l} X^l_{(0)s} + v^{(0)}_{s,l} X^l_{(0)r} - v^{(0),p}_p X^{(0)}_{rs} , \tag{9.7}$$

In (9.7) the dot denotes differentiation with respect to time, and the index (0) means that a function is evaluated at the instant $t = 0$. Since any in-

---

[1] Other requirements for a physically acceptable definition of stress rate have been discussed by W. PRAGER.

stant may be the initial one, by identifying the initial instant with the actual one and supposing now $\bar{C}$ different from C^*, from (9.7) one deduces

$$\dot{X}_{rs} - v_r^{/l} X_{ls} - v_s^{/l} X_{lr} + v_p^{/p} X_{rs} = \frac{1}{D} \dot{Y}^{lm} x_{r,l} x_{s,m} . \tag{9.8}$$

From (9.8) it follows that the $\dot{Y}_{rs}$ are zero for any rigid motion if and only if the left-hand members of (9.8) are zero for such a motion. In effect, the first members of (9.8) may be interpreted as propagation velocities of stress.

From the above considerations it follows that in (9.4), (9.5) it is appropriate to take

$$\frac{\delta \xi_{rs}}{\delta t} = \dot{\xi}_{rs} - v^r/_l \xi_{ls} - v^s/_l \xi_{lr} + \xi_{rs} v^p/_p . \tag{9.9}$$

The last term of the second member of (9.9) may be suppressed, since it is zero for any rigid motion and one may, generally, be absorbed into the second members of (9.4), (9.5).

Therefore, the basic equations of hypo-elasticity are the constitutive equations

$$\frac{\partial \xi_{rs}}{\partial t} + \xi_{rs}/_p v^p - \xi_{ls} v^r/_l - \xi_{lr} v^s/_l + \xi_{rs} v^p/_p = A^{rspq} d_{pq} , \tag{9.10}$$

the dynamical equations

$$2 \mu \xi_{rs}/_s = k (F^r - \dot{v}^r) \tag{9.11}$$

and the equation of mass

$$\dot{k} + k v^p/_p = 0 . \tag{9.12}$$

Together with (9.10), (9.11), (9.12) the initial and boundary conditions are to be borne in mind.

3. The property of isotropy for a body must be considered equivalent to the property that the tensor A^{rspq} be isotropic. C. TRUESDELL [1] proposed an expression of the tensor A^{rspq} for isotropic bodies. Later it was demonstrated (RIVLIN and ERICKSEN) that the expression of A^{rspq} established by TRUESDELL is fully general, if an adequate definition of isotropic function is given. Then it is possible to demonstrate that the most general expression of the tensor[1] $A^{rspq} d_{pq}$ is the following:

$$A^{rspq} d_{pq} = (g_0 d^p_{\ p} + M g_3 + N g_7) \delta^{rs} + g_1 d^{rs} + (g_2 d^p_{\ p} + M g_6 + N g_{10}) \xi^{rs} +$$

$$+ \frac{1}{2} g_4 (d^{rl} \xi_l^{\cdot s} + d^{ls} \xi_l^{\cdot r}) + (g_5 d^p_{\ p} + M g_9 + N g_{11}) \xi^{rl} \xi_l^{\cdot s} +$$

$$+ \frac{1}{2} g_8 (d^r_{\ t} \xi^{tp} \xi_p^{\cdot s} + \xi^r_{\cdot t} \xi^t_{\ p} d^{ps}) , \tag{9.13}$$

---

[1] Clearly, expressions like (3.8), (3.14) are particular cases of the general dependence (9.13) of an isotropic tensor on its arguments.

where coefficients g_q are dimensionless analytic functions of the three principal invariants of ξ^{rs}:

$$I\,(\xi) = \xi^i_{\cdot i}\,, \qquad II = \frac{1}{2}\,\delta^{ij}_{kl}\,\xi^k_{\cdot i}\,\xi^l_{\cdot j}\,, \qquad III = \frac{1}{6}\,\delta^{ijk}_{lmn}\,\xi^l_{\cdot i}\,\xi^m_{\cdot j}\,\xi^n_{\cdot k} \qquad (9.14)$$

and

$$M = \xi^i_{\cdot j}\,d^j_{\cdot i}\,, \qquad N = \xi^i_{\cdot j}\,\xi^j_{\cdot k}\,\xi^k_{\cdot i}\,. \qquad (9.15)$$

If, in particular, one supposes

$$g_0 = \frac{\lambda}{2\,\mu}\,, \qquad g_1 = 1\,, \qquad g_2 = \cdots = g_{11} = 0\,, \qquad (9.16)$$

the body is called *hypo-elastic of grade zero*, and (9.10) becomes

$$\frac{\delta\,\xi^{rs}}{\delta\,t} = \frac{\lambda}{2\,\mu}\,\delta^{rs} d^p_p + d^{rs}\,. \qquad (9.17)$$

Generally, supposing that g_i are polynomials in I, II, III such that the A^{rspq} are of grade n in ξ^{lm}, the body is said *hypo-elastic of grade n*. Hypo-elasticity, based on the set of equations (9.10), (9.11), (9.12), has wider possibilities than the theory of finite elastic deformations, based on the assumption of the existence of a natural state, and it may be that bodies which cannot be studied by the classical theory of elasticity may be treated by hypo-elasticity theory.

Now, instead of the difficult problem of determining the elastic potential one has that of determining the tensor A^{rspq}.

This is a more difficult problem, since it is necessary that experiment and mathematics give the structure of every invariant function that is present in (9.13). This problem certainly should be simplified by using the thermodynamic laws and by studying hypo-elasticity from an energetic point of view. Much remains to be done in this field, but some results have already been obtained. For example, BERNSTEIN and ERICKSEN have characterized some cases where the density of the work done by the internal forces does not depend on the stress history but only on the initial and final states.

VERMA [3] has shown that for hypo-elastic bodies of grade zero, the internal energy is the same function of stress as in the linear elasticity theory.

For other results concerning several problems of hypo-elasticity see also GREEN [2, 3], VERMA [1, 2], ERICKSEN [2], THOMAS.

§ 3. Hypo-elastic equilibrium

1. In the statics of classical elasticity the problem of determining the stress equilibrated by the external forces and deriving from a real displacement is fundamental. The corresponding problem of hypo-elasticity is evidently that of finding a solution of equations (9.10), (9.11),

(9.12) representing a motion without acceleration. It is to be presumed that such a solution exists and is determined by initial and boundary conditions. The static problem of hypo-elasticity is therefore expressed by equations (9.10), (9.12), by equations

$$2\mu\, \xi^{rs}/_s = kF^r \,,\tag{9.18}$$

$$\frac{\partial v^r}{\partial t} + v^r/_p v^p = 0 \,,\tag{9.19}$$

and by initial and boundary conditions. If one expresses a fundamental element of deformation as a function of the time, inverting such a functional relation gives the time as a function of the deformation. By eliminating t it is possible to establish relations among stress and deformation. Theorems of existence and uniqueness for equations (9.10), (9.12), (9.18), (9.19) and their initial and boundary conditions are not yet established.

2. If body forces are absent $[F^r = 0]$ and if the v_i are supposed to be known linear functions of the x_j [homogeneous deformation],

$$v_i = a_{ij} x^j + b_i\tag{9.20}$$

where a_{ij}, b_i are functions of the time alone, a uniform stress that is independent of x^i satisfies equations (9.10), (9.18). In fact, in this case (9.18) becomes an identity, while (9.10) becomes

$$\frac{d\,\xi^{rs}}{dt} = \Psi^{rs}(\xi^{ih}, t).\tag{9.21}$$

In particular, equations (9.21) are linear in the case of hypo-elastic bodies of grade zero and of grade one.

If a_{ij}, b_i are analytic functions of t, equations (9.21) admit a unique solution independent of x^p which corresponds to an assigned homogeneous stress for $t = 0$.

It is easy to recognize that the most general solution of (9.19) of the kind (9.20) is determined by the equations

$$\left.\begin{array}{l}(\delta_i^l + A_i^{\,l}t)\, a_{lm} = A_{im} \,,\\[4pt] (\delta_i^l + A_i^{\,l}t)\, b_l = B_i \,,\end{array}\right\}\tag{9.22}$$

where A_{il}, B_i are arbitrary constants whose meaning is, clearly,

$$A_{im} = a_{im}(0) \,, \qquad B_i = b_i(0) \,.\tag{9.23}$$

The solution corresponding to (9.20) is valid in any interval δ_t of the time which contains the initial instant $t = 0$ and none of the possible values of t which annul the determinant

$$\varDelta(t) = \|\delta^{il} + A^{il}t\| . \tag{9.24}$$

The interval δ_t may be infinitely large.

The case of a stress compatible with a homogeneous deformation has been studied by C. TRUESDELL [4, 5, 6, 7], who has demonstrated a theorem of existence and uniqueness and has examined some interesting cases, such as that of simple extension, of hydrostatic pressure, of simple shear, etc.

§ 4. Lagrangean expression of the hypo-elasticity equations

It may be useful to express the general equations of hypo-elasticity in a Lagrangean form, at least with the aim of studying the question of their solutions.

Let us assume as reference state C^* the initial one C_0 and call y_i the co-ordinates of the points of C_0. In analogy to (2.10) let us suppose

$$\eta^{rs} = \frac{Y^{rs}}{2\mu} , \qquad \xi^{rs} = \frac{1}{D}\,\eta_{lm}\,x^{r,l}\,x^{s,m} . \tag{9.25}$$

From (9.25, 2) follows

$$\dot{\xi}^{rs} = -\,\frac{\dot{D}}{D^2}\,\eta_{lm}\,x^{r,l}\,x^{s,m} + \frac{1}{D}\,\big[\dot{\eta}_{lm}\,x^{r,l}\,x^{s,m} + \eta_{lm}\,(\dot{x}^{r,l}\,x^{s,m} + x^{r,l}\,\dot{x}^{s,m})\big], \tag{9.26}$$

while

$$\left.\begin{array}{c} v_{r/s} = v_{r,m}\,y^m{}/_s = \dot{x}_{r,m}\,\dfrac{C_s{}^{\cdot m}}{D} , \\[2mm] \dot{x}_{r,s} = v_{r,m}\,x^m{}_{,s} \end{array}\right\} \tag{9.27}$$

and, therefore,

$$d_{rs} = \frac{1}{2D}\,(\dot{x}_{r,p}\,C_s{}^{\cdot p} + \dot{x}_{s,p}\,C_r{}^{\cdot p}) . \tag{9.28}$$

From (9.25, 2), (9.26), (9.27), the following Lagrangean expression of equations (9.10) is deduced:

$$\dot{\eta}^{lm}\,x^r{}_{,l}\,x^s{}_{,m} = \frac{A_*^{rspq}}{2}\,(\dot{x}_{p,m}\,C_q{}^{\cdot m} + \dot{x}_{q,m}\,C_p{}^{\cdot m}) , \tag{9.29}$$

where A_*^{rspq} are the expressions obtained from A^{rspq} by substituting for ξ^{rs} their expressions (9.25, 2). The same result may be established starting from (9.8). With (9.29) are to be associated the Lagrangean equations corresponding to (9.11), (9.12), which are, plainly,

$$(x_{r,l}\,\eta^{ls})_{,s} = k^*(F_r - \ddot{x}_r) , \tag{9.30}$$

$$kD = k^* . \tag{9.31}$$

The ten equations (9.29), (9.30), (9.31) contain the ten unknown quantities η^{lm}, x_r, k. In the static case the $\dot{x}_r$ are independent of t and the $\ddot{x}_r$ are equal to zero. As a consequence of the Lagrangean expression of the hypo-elasticity equations it is possible to demonstrate that every isotropic elastic body is a special hypo-elastic one. This has been recognized and demonstrated by W. NOLL; it may easily be deduced by differentiation of (2.10), keeping in mind (9.29).

§ 5. On the solution of the equations of hypo-elasticity in the static case for the body of grade zero

The Lagrangean expression (9.29) of the constitutive equations is particularly useful for studying the problem of finding the solutions in the static case of hypo-elastic bodies of grade zero. Under such a hypothesis the x_r are linear functions of t. That is,

$$\left. \begin{aligned} x_r &= \alpha_r(y_i)\, t + y_r\,, \\ x_{r,s} &= \alpha_{r,s} t + \delta_{rs}\,, \qquad \dot{x}_{r,s} = \alpha_{r,s}\,. \end{aligned} \right\} \qquad (9.32)$$

From (9.17), (9.28), (9.32) it follows that in the case of bodies of grade zero (9.29) becomes

$$\dot{\eta}^{lm} x_{r,l} x_{s,m} = \frac{\lambda}{2\mu}\, \delta_{rs}\, \dot{x}_{p,q}\, C^{pq} + \frac{1}{2}\, (\dot{x}_{r,p}\, C_s^{\cdot p} + \dot{x}_{s,p}\, C_r^{\cdot p})\,. \qquad (9.33)$$

From (9.33) follows

$$\dot{\eta}^{rs} = \frac{C^{lr} C^{ms}}{2 D^2} \left[\frac{\lambda}{\mu}\, \delta_{lm}\, \dot{x}_{p,q}\, C^{pq} + \dot{x}_{l,p}\, C_m^{\cdot p} + \dot{x}_{m,p}\, C_l^{\cdot p} \right]\,. \qquad (9.34)$$

The right-hand member of (9.34), in view of (9.32), is a known function of t, and therefore

$$\eta^{rs} - \eta_0^{rs} = \frac{1}{2} \int_0^t \left(\frac{C^{lr} C^{ms}}{D^2} \right)_{t=\tau} \left[\frac{\lambda}{\mu}\, \delta_{lm}\, \dot{x}_{p,q}\, C^{pq} + \dot{x}_{l,p}\, C_m^{\cdot p} + \dot{x}_{m,p}\, C_l^{\cdot p} \right]_{t=\tau} d\tau\,. \qquad (9.35)$$

The η^{rs} are determined by calculation of the integrals of the right-hand members of (9.35). Then (9.25, 2) gives ξ^{rs}. If one wishes to have ξ^{rs} as a function of x_i, it is necessary to deduce t as a function of x_i from (9.32). In effect, (9.35) gives the η^{rs} as functions of $\alpha_{r,s}(y_i)$ and t. In the static case equations (9.30) become

$$(x_{r,l}\, \eta^{ls})_{,s} = k^* F_r \qquad (9.36)$$

and together with the boundary conditions represent the equations that determine $\alpha_r(y_i)$, *if a static solution exists.*

2. To determine static solutions, where v_r are specified functions of x_i and of t, that is, $v_r = L_r(x_i; t)$, and the acceleration calculated from v_r vanishes, we must have

$$\frac{\partial L_r}{\partial t} + L_{r/h} L^h = 0 .$$ (9.37)

In this case the $\alpha_r(y_i)$ are expressed by

$$\alpha_r(y_i) = L_r(y_i, 0) .$$ (9.38)

In the special case of a *homogeneous deformation* (9.20), (9.22), (9.23) are valid, and $\alpha_r(y_i)$ cannot have an expression different from

$$\alpha_r(y_i) = A_{rl} y^l + B_r .$$ (9.39)

Now if body forces are absent $[F_r = 0]$, (9.36) admits, plainly, a solution with all the η^{rs} equal to constants.

§ 6. Some considerations concerning the solution of the general problem of hypo-elastic statics

1. If a hypo-elastic body is not of grade zero, the right-hand member of (9.29) depends on η^{rs}, and the solution of the set of equations (9.29), (9.30) is generally very difficult, even in the static case, where the $\ddot{x}_r$ are all equal to zero. Theorems of existence and uniqueness are in general still to be established. However, it may be observed that if a solution of the static problem exists and if it is possible to express it by means of a power series in t, then one may construct the series by a procedure formally analogous to the one of Chapter IV.

Let us observe, in effect, that from (9.29) follows

$$\dot{\eta}^{rs} = F^{rs}(x_{p,q}, \dot{x}_{l,m}, \eta^{nh}) .$$ (9.40)

In the static case equations (9.32) are valid, and the differential equations (9.40), (9.36) become

$$\dot{\eta}^{rs} = f^{rs}[\alpha_{p,q}(y_i), \eta^{lm}; t] ,$$ (9.41)

$$\eta^{rs}_{,s} + t(\alpha^{rl} \eta_{ls})^{,s} = k^* F^r .$$ (9.42)

Then, a static solution exists if and only if there are nine functions $\alpha_r(y_i), \eta^{rs}(y_i, t)$ which verify (9.41), (9.42) and the initial and boundary conditions.

From (9.41) one may formally deduce the successive derivatives of η^{rs} for $t = 0$, obtaining

$$(\ddot{\eta}^{rs})_{t=0} = \left(\frac{\partial f^{rs}}{\partial t}\right)_{t=0} + \sum_{l,m}\left(\frac{\partial f^{rs}}{\partial \eta^{lm}} f^{lm}\right)_{t=0}, \quad \text{etc.}$$ (9.43)

On the other hand, from (9.42) follows

$$[\eta^{rs}_{,s}]_{t=0} = k^* F^r(0) , \qquad [\dot\eta^{rs}_{,s}]_{t=0} = k^* \dot F^r(0) - (\alpha^{r,l} \eta^{(0)}_{ls})_{,s} , \qquad (9.44)$$

and, in general,

$$\eta^{rs(n)}_{(0),s} = h^* F^{r(n)}_{(0)} - n(\alpha^{r,l} \eta^{(0)(n-1)}_{ls})_{,s} . \qquad (9.45)$$

(9.41), (9.43) expressed in terms of the derivatives of every order with respect to t at $t=0$ must be compatible[1] with (9.44), (9.45). That represents a strong condition on $\alpha_r(y_i)$. In effect (9.41), (9.43) determine all derivatives of η^{rs} for $t=0$ as functions of $\alpha_{p,q}$ and $\eta^{lm}_{(0)}$. Introducing such derivatives in (9.44), (9.45), we see that a solution of (9.41), (9.42) exists and is expressible in power series of t only if there is a set of $\alpha_{p,q}(y_i)$, $\eta^{lm}_{(0)}(y_i)$ which satisfy (9.44), (9.45) and the initial and boundary conditions. In correspondence to such set of $\alpha_{p,q}(y_i)\,\eta^{lm}_{(0)}(y_i)$, (9.41), (9.43) give all derivatives of η^{rs} for $t=0$, and it is possible to obtain formally a Taylor's series which, except for the problem of convergence, represents the solution of the static problem. Plainly, the analogous Taylor's series for ξ^{rs} may be expressed by calculation of successive derivatives of ξ^{rs} for $t=0$, according to (9.25, 2), (9.26)

2. It is evident that (9.42) is satisfied by every set of $\alpha_{p,q}$ and η^{rs} independent of y_i, if body forces are absent. Hence it is easy in this way to get any solution representing a homogeneous deformation and a uniform stress which is expressible by power series. Rather, in such a case, on account of the independence of $v_{i/h}$ from x_l it may be convenient to apply the method directly to ξ^{rs}. For example, let us consider the case of simple shear, studied by TRUESDELL in the papers cited, represented by the velocities

$$v_1 = 2hx_2 , \qquad v_2 = v_3 = 0 , \qquad (9.46)$$

where h is a constant. Taking Taylor's series only up to and including the term of fifth degree in t, one finds, for a *hypo-elastic body of grade one*, and if ξ^{11}_0, ξ^{12}_0 are zero,

$$\xi^{22} = \xi^{33} = \xi^{13} = \xi^{23} = 0 , \qquad (9.47)$$

$$\xi^{11} = \tau^2 + \frac{a_1}{6}\tau^4 ,$$
$$\xi^{12} = \tau + \frac{a_1}{3}\tau^3 + \frac{a_1^2}{30}\tau^5 , \qquad (9.48)$$

where a_1 is a constant and $\tau = ht$.

---

[1] Clearly, a necessary condition that a solution exists which admits derivatives for $t=0$ with respect to t up to order m is that (9.41), (9.43) be compatible with (9.44), (9.45), for $n = 1, 2, \ldots, m$.

Let us suppose $a_1 < 0$. ξ^{12} has a relative maximum when

$$\tau = \tau^* = \sqrt{\frac{-3+\sqrt{3}}{a_1}} \approx \frac{1{,}13}{\sqrt{-a_1}}. \qquad (9.49)$$

By solution of the equations of hypo-elasticity (TRUESDELL [5]) one finds, instead,

$$\tau^* = \frac{\pi}{2\sqrt{-2a_1}} \approx \frac{1{,}11}{\sqrt{-a_1}}. \qquad (9.50)$$

It is evident that the values (9.49), (9.50) are close numerically. In other words, even assuming only a few terms of Taylor's series it is possible, in some cases, to obtain a numerical value of shear such that the body will collapse or yield at or before this strain.

Specifically, from (9.48), (9.49) there results for the critical stress the value

$$(\xi^{12})_{\max} = \sqrt{\frac{-3+\sqrt{3}}{a_1}} \; \frac{2\sqrt{3}+6}{15} \approx \frac{0{,}71}{\sqrt{-a_1}}, \qquad (9.51)$$

while the exact solution gives

$$(\xi^{12})_{\max} = \frac{1}{\sqrt{-2a_1}} \approx \frac{0{,}71}{\sqrt{-a_1}}. \qquad (9.52)$$

If the body is of grade zero, $a_1 = 0$. In this case (9.48) represents [for $a_1 = 0$] the exact solution, and it is very interesting to observe that the classical law of proportionality between shear strain and shear stress remains true, but the presence of a normal tension on the planes perpendicular to the y_3 axis is necessary.

Chapter X

Asymmetric Elasticity

§ 1. Preliminaries

In this chapter I consider elastic equilibrium under the hypothesis that the stress components need not be symmetric; that is, $X_{rs} \neq X_{sr}$ [and $Y_{rs} \neq Y_{sr}$]. The best known books on the theory of elasticity do not in general consider this case, and only a few papers are devoted to it. SOMIGLIANA has established some general equations for slightly deformable bodies. Later BODAZEWSKI has considered the problem, also applying it to hydrodynamics. However, the results of these authors are based on an expression of the work done by the internal contact forces [SOMIGLIANA] or on linear stress-strain relations [BODAZEWSKI] which do not seem acceptable. The cause of asymmetry of X_{rs} may be the presence of body moments. This means that the body forces acting upon an infinitesimal element are reducible to a force applied at a point of the element and a couple, as happens, for example, in the presence of magnetic forces. This case is considered the most interesting one.

However, even when there are no body moments, it may happen [REISSNER] that the problem considered implies asymmetry for the stress components, at least in a certain part of the body. In this case the theory with symmetric X_{rs} necessarily gives solutions having singularities [multi-valued functions, infinities], and it is possible that often plastic yield is reached just for this reason.

Therefore, a theory in which the X_{rs} are asymmetric even in the absence of body moments may be interesting, but it is impossible to exclude the presence of *surface moments*. That is, it must necessarily be admitted that the set of internal contact forces through a surface element are generally reducible to a force applied at a point of the element and to a couple whose moment I shall name the *surface moment*. In such manner, problems which, according to the usual theory, have only solutions with physically implausible singularities attain regular solutions.

I think that the difficulty [or impossibility] of realizing external surface moments need not imply that the existence of internal surface moments be absurd, as some authors think (VOIGT [1]), if one considers the body elements very small but not vanishing. After all, it is not even easy to show in what manner the set of external surface forces typically hypothecated in the usual theory may be realized.

Therefore, I wish to set forth the theory of elastic bodies with asymmetric stress components, assuming the presence of surface moments. I intend to consider the case of large displacements as being interesting not only in itself but also to study small deformations: in fact, if one establishes directly a linear theory by a procedure analogous to the one used in the symmetric case, an indeterminancy remains in the structure of the thermodynamic potential [and, therefore, in the stress-strain relations] which may be difficult to remove if the linear theory is not deduced from the one of large deformations as a first approximation.

The stress components are derived from the knowledge not only of the thermodynamic potential but also of a parameter. That is, there is a formal analogy with the case of the elasticity of incompressible bodies (SIGNORINI [8, 9]), but with substantial differences. Considering the case of isotropic slightly deformable bodies, and supposing acceptable the idea that surface moments are zero for any irrotational displacement starting from a natural state of equilibrium, it is possible to show that the above parameter is zero and, then, that the stress is derived only from a potential function.

It is evident that the general equations which will be established may be extended to the dynamical case.

For more explicit details on the subject see GRIOLI [11].

§ 2. Basic Eulerian equations

1. Let us suppose that the body forces acting upon the element dC of the actual equilibrium state are equivalent to the force $k\boldsymbol{F}dC$ applied at a point of dC and a couple of moment $k\boldsymbol{M}dC$, while the external surface forces acting upon an element $d\Sigma$ of the boundary are reducible to the force $\boldsymbol{f}d\Sigma$ applied at a point of $d\Sigma$ and a couple of moment $\boldsymbol{m}d\Sigma$.

Further, supposing that the internal contact forces are reducible in an analogous manner, let these forces be characterized by the vectors $\boldsymbol{\Phi}_v$, $\boldsymbol{\Psi}_v$ where $\boldsymbol{\Phi}_v$ has the usual meaning [see Chapter II, § 1], while $\boldsymbol{\Psi}_v$ characterizes the moment of the couple, and for it the conventions already stated for $\vec{\boldsymbol{\Phi}}_v$ are valid.

The basic equations of equilibrium corresponding to an arbitrary portion c internal to C of boundary σ and inward normal $\boldsymbol{n}$ are

$$\int_\sigma \boldsymbol{\Phi}_n d\sigma + \int_c k\boldsymbol{F}dC = 0 , \tag{10.1}$$

$$\int_\sigma OP \wedge \boldsymbol{\Phi}_n d\sigma + \int_c OP \wedge k\boldsymbol{F}dC + \int_\sigma \boldsymbol{\Psi}_n d\sigma + \int_c k\boldsymbol{M}dC = 0 . \tag{10.2}$$

From (10.1), (10.2) easily follows

$$\boldsymbol{\Psi}_v = \boldsymbol{\Psi}_s v^s , \tag{10.3}$$

where the meaning of the vector $\boldsymbol{\Psi}_s$ is analogous to that of $\boldsymbol{\Phi}_s$ but with reference to surface moments.

From (10.1), (10.2), (10.3) the Eulerian equilibrium equations follow:

$$\left.\begin{aligned} \boldsymbol{\Phi}^s_{,s} &= k\boldsymbol{F} && \text{(in } C\text{) ,} \\ \boldsymbol{\Phi}^s N_s &= \boldsymbol{f} && \text{(on } \Sigma\text{) ,} \end{aligned}\right\} \tag{10.4}$$

$$\left.\begin{aligned} \boldsymbol{\Psi}^s_{,s} &= -\boldsymbol{c}_s \times \boldsymbol{\Phi}^s + k\boldsymbol{M} && \text{(in } C\text{) ,} \\ \boldsymbol{\Psi}^s N_s &= \boldsymbol{m} && \text{(on } \Sigma\text{) .} \end{aligned}\right\} \tag{10.5}$$

2. Bearing in mind (2.6), let us suppose

$$X_{rs} = \boldsymbol{c}_r \cdot \boldsymbol{\Phi}_s , \qquad \Psi_{rs} = \boldsymbol{c}_r \cdot \boldsymbol{\Psi}_s . \tag{10.6}$$

For an infinitesimal displacement of the body, starting from the actual state, of components δu_r, let $\delta'\omega$ be the local rotation. Its components are

$$\delta'\omega_r = \frac{1}{2}\left[(\delta u_{r+2})_{/r+1} - (\delta u_{r+1})_{/r+2}\right] , \tag{10.7}$$

where, as usual, an oblique dash (/) before an index s denotes differentiation with respect to x_s.

From (10.4, 1), (10.5, 1) it follows that the work done by the internal surface forces is

$$\delta \mathfrak{L}^{(i)} = - \int_C \boldsymbol{\Phi}^s_{,s} \cdot \delta P dC - \int_\Sigma \boldsymbol{f} \cdot \delta P d\Sigma - \int_C \boldsymbol{\Psi}^s_{,s} \cdot \delta' \omega dC -$$
$$- \int_C \boldsymbol{c}_s \times \boldsymbol{\Phi}^s \cdot \delta' \omega dC - \int_\Sigma \boldsymbol{m} \cdot \delta' \omega d\Sigma , \tag{10.8}$$

which by (10.4, 2), (10.5, 2) becomes

$$\delta \mathfrak{L}^{(i)} = \int_C \delta l^{(i)} dC , \tag{10.9}$$

where

$$\delta l^{(i)} = X^{rs} (\delta u_r)_{/s} + \Psi^{rs} (\delta' \omega_r)_{/s} + \sum_r (X_{r+1\,r+2} - X_{r+2\,r+1}) \delta' \omega_r . \tag{10.10}$$

It is easy to show that the direct dependence of $\delta l^{(i)}$ on $\delta' \omega_r$ is apparent[1]. In fact, from (10.7), (10.10) follows

$$\delta l^{(i)} = \sum_r \{ X_{rr} (\delta u_r)_{/r} + X_{r+1\,r+2} (\delta u_{r+1})_{/r+2} +$$
$$+ X_{r+2\,r+1} (\delta u_{r+2})_{/r+1} + \frac{1}{2} (X_{r+1\,r+2} - X_{r+2\,r+1}) [(\delta u_{r+2})_{/r+1} -$$
$$- (\delta u_{r+1})_{/r+2}] \} + \Psi^{rs} (\delta' \omega_r)_{/s} , \tag{10.11}$$

and then

$$\delta l^{(i)} = \sum_r \left\{ X_{rr} (\delta u_r)_{/r} + \frac{1}{2} (X_{r+1\,r+2} + X_{r+2\,r+1}) [(\delta u_{r+1})_{/r+2} + \right.$$
$$\left. + (\delta u_{r+2})_{/r+1}] \right\} + \Psi^{rs} (\delta' \omega_r)_{/s} . \tag{10.12}$$

Putting

$$\xi_{rs} = \frac{X_{rs} + X_{sr}}{2} = \xi_{sr} , \tag{10.13}$$

one has

$$\delta l^{(i)} = \xi^{rs} \delta' e_{rs} + \Psi^{rs} (\delta' \omega_r)_{/s} , \tag{10.14}$$

with

$$\delta' e_{rs} = \frac{1}{2} [(\delta u_r)_{/s} + (\delta u_s)_{/r}] . \tag{10.15}$$

Observing (10.14), one recognizes that if $\boldsymbol{m}$ and Ψ^{rs} are supposed equal to zero [surface moments absent], $\delta l^{(i)}$ does not depend on the local rotation, not even if X^{rs} is asymmetric from the presence of body moments.

---

[1] Instead, the expression of $\delta l^{(i)}$ corresponding to the case of small deformations in [SOMIGLIANA] depends essentially on $X_{rs} - X_{sr}$ and $\delta' \omega_r$. Then $\delta l^{(i)}$ is not zero for a rigid displacement.

§ 3. Basic Lagrangean equations

1. From (10.1), (10.2) follow

$$\int_\sigma X_{rs} n^s d\sigma + \int_c k F_r dC = 0 , \qquad (10.16)$$

$$\int_\sigma (x_{r+1} X_{r+2s} - x_{r+2} X_{r+1s}) n^s d\sigma + \int_c (x_{r+1} F_{r+2} - x_{r+2} F_{r+1} +$$
$$+ M_r) k dC + \int_\sigma \Psi_{rs} n^s d\sigma = 0 . \qquad (10.17)$$

Analogously to what has been done in Chapter II, § 2, let us suppose

$$m^* d\Sigma^* = m d\Sigma , \qquad (10.18)$$

$$\Psi_{rs} = \frac{1}{D} \varphi^{lm} x_{r,l} x_{s,m} . \qquad (10.19)$$

Bearing in mind (10.18), (2.10) and (10.19), we see that (10.17) becomes

$$\int_{\sigma^*} \frac{1}{D} x_{s,m} (x_{r+1} x_{r+2,l} - x_{r+2} x_{r+1,l}) Y^{lm} C^{st} n_t^* d\sigma^* + \int_{C^*} (x_{r+1} F_{r+2} -$$
$$- x_{r+2} F_{r+1} + M_r) k^* dC^* + \int_{\sigma^*} \frac{1}{D} \varphi^{lm} x_{r,l} x_{s,m} C^{st} n_t^* d\sigma^* = 0 . \quad (10.20)$$

From (10.20) follows

$$\int_{\sigma^*} Y^{lm} (x_{r+1} x_{r+2,l} - x_{r+2} x_{r+1,l}) n_m^* d\sigma^* + \int_{C^*} (x_{r+1} F_{r+2} -$$
$$- x_{r+2} F_{r+1} + M_r) k^* dC^* + \int_{\sigma^*} \varphi^{lm} x_{r,l} n_m^* d\sigma^* = 0 , \qquad (10.21)$$

from which, bearing in mind (2.12), one deduces

$$\int_{C^*} Y^{lm} (x_{r+2,l} x_{r+1,m} - x_{r+1,l} x_{r+2,m}) dC^* + \int_{C^*} (\varphi^{lm} x_{r,l})_{,m} dC^* -$$
$$- \int_{C^*} M_r^* k^* dC^* = 0 . \qquad (10.22)$$

By the arbitrariness of c^* and the fact that the functions are independent of the field of integration, from (10.22) follows, almost everywhere,

$$(\varphi^{lm} x_{r,l})_{,m} = (x_{r+1,l} x_{r+2,m} - x_{r+2,l} x_{r+1,m}) Y^{lm} + k^* M_r . \quad (10.23)$$

Putting

$$\lambda_{rm} = \varphi_{lm} x_r^{,l} \qquad (10.24)$$

and again writing (2.12), one deduces, then, that the Lagrangean equilibrium equations are

$$(Y^{lm} x_{r,l})_{,m} = k^* F_r ,$$ (10.25)

$$\lambda_{rm}^{\cdot m} = (x_{r+1,l} x_{r+2,m} - x_{r+1,m} x_{r+2,l}) Y^{lm} + k^* M_r ,$$ (10.26)

with which are to be associated the boundary equations

$$\left. \begin{array}{l} Y^{lm} x_{r,l} N_m^* = f_r^* , \\ \lambda_{rm} N_*^m = m_r^* . \end{array} \right\}$$ (10.27)

2. Putting

$$T_{lm} = T_{ml} = \frac{Y_{lm} + Y_{ml}}{2} ,$$ (10.28)

from (2.10), (10.13) we show that

$$\xi_{rs} = \frac{1}{D} T^{lm} x_{r,l} x_{s,m} .$$ (10.29)

Since

$$\left. \begin{array}{l} (\delta u_r)_{|s} = \dfrac{1}{D} C_{sl} (\delta u_r)^{\cdot l} , \\ (\delta' \omega_r)_{|s} = \dfrac{1}{D} C_{sl} (\delta' \omega_r)^{\cdot l} , \end{array} \right\}$$ (10.30)

from (10.14), (10.19), (10.29), (10.30) follows

$$\delta l^{(i)} = \frac{1}{D^2} x_{r,l} x_{s,m} \left\{ \frac{1}{2} T^{lm} [C^{sq} (\delta u^r)_{,q} + C^{rq} (\delta u^s)_{,q}] + \right.$$
$$\left. + \varphi^{lm} C^{sq} (\delta' \omega^r)_{,q} \right\} .$$ (10.31)

Then setting $\delta l^{(i)} dC = \delta^* l^{(i)} dC^*$, from (10.31) one deduces

$$\delta^* l^{(i)} = T^{lm} (\delta u_r)_{,m} x_{,l}^r + \varphi^{lm} (\delta' \omega_r)_{,m} x_{,l}^r ,$$ (10.32)

which according to (1.4) may be written

$$\delta^* l^{(i)} = T^{lm} \frac{\delta b_{lm}}{2} + \lambda^{lm} (\delta' \omega_l)_{,m} .$$ (10.33)

3. In the case of finite deformations it is not evident of which functions of y_i the quantities $\delta' \omega_r$ are the variations. Because of this inconvenience the elegant thermodynamic procedure, which in the symmetric
case allows one to derive Y_{rs} from the thermodynamic potential, is not
directly applicable to expression (10.33) of $\delta^* l^{(i)}$, and this expression
must be transformed. To begin with I observe that

$$\delta(u_{i,s}) = (\delta u_i)_{,s} , \qquad \delta(u_{i,sm}) = (\delta u_i)_{,sm} ,$$ (10.34)

while if $F(x)$ is a function which depends on y_r only through $x_{i,h}$, one has

$$
\begin{aligned}
\delta(F_{,s}) &= \delta\left[\sum_{i,h}\frac{\partial F}{\partial x_{i,h}}u_{i,hs}\right] = \\
&= \sum_{i,h}\left[\frac{\partial F}{\partial x_{i,h}}(\delta u_{i,h})_{,s} + \sum_{pq}\frac{\partial^2 F}{\partial x_{i,h}\partial x_{p,q}}u_{i,hs}\,\delta(u_{p,q})\right], \\
(\delta F)_{,s} &= \left[\sum_{i,h}\frac{\partial F}{\partial x_{i,h}}\delta(u_{i,h})\right]_{,s} = \\
&= \sum_{i,h}\left[\frac{\partial F}{\partial x_{i,h}}(\delta u_{i,h})_{,s} + \sum_{pq}\frac{\partial^2 F}{\partial x_{i,h}\partial x_{p,q}}u_{p,qs}\,\delta(u_{i,h})\right].
\end{aligned} \right\} \quad (10.35)
$$

Then

$$(\delta F)_{,s} = \delta(F_{,s}) . \tag{10.36}$$

By (10.30) it is seen that $\delta'\omega_r$ may be put in the Lagrangean form

$$\delta'\omega_r = \frac{1}{2D}\left[(\delta u_{r+2})_{,m}C_{r+1}^{\cdot m} - (\delta u_{r+1})_{,m}C_{r+2}^{\cdot m}\right]. \tag{10.37}$$

From (10.34), (10.35), (10.36), (10.37) follows

$$
\begin{aligned}
(\delta'\omega_r)_{,s} = \frac{1}{2}\sum_{pqm}&\left\{\left[\delta(u_{r+2,m})\frac{\partial}{\partial x_{p,q}}\left(\frac{C_{r+1\,m}}{D}\right) - \delta(u_{r+1\,m})\frac{\partial}{\partial x_{p,q}}\left(\frac{C_{r+2\,m}}{D}\right)\right]u_{p,sq}\right. \\
&-\left.\left[u_{r+2,sm}\frac{\partial}{\partial x_{p,q}}\left(\frac{C_{r+1\,m}}{D}\right) - u_{r+1,sm}\frac{\partial}{\partial x_{p,q}}\left(\frac{C_{r+2\,m}}{D}\right)\right]\delta(u_{p,q})\right\} + \\
&+ \delta\left[\frac{1}{2D}\sum_m(u_{r+2,sm}C_{r+1\,m} - u_{r+1,sm}C_{r+2\,m})\right]. \tag{10.38}
\end{aligned}
$$

Then with the notations

$$\mu_{rs} = \frac{1}{2D}\left[u_{r+2,sm}C_{r+1}^{\cdot m} - u_{r+1,sm}C_{r+2}^{\cdot m}\right], \tag{10.39}$$

$$
\begin{aligned}
M_{rs} &= \frac{1}{2}\sum_{pqm}\left[u_{p+1,qm}\frac{\partial}{\partial x_{r,s}}\left(\frac{C_{p+2\,m}}{D}\right) - u_{p+2,qm}\frac{\partial}{\partial x_{r,s}}\left(\frac{C_{p+1\,m}}{D}\right)\right]\lambda_{pq}, \\
N_{rs} &= \frac{1}{2}\sum_{pqm}\left[\lambda_{r+1\,m}\frac{\partial}{\partial x_{p,q}}\left(\frac{C_{r+2\,s}}{D}\right) - \lambda_{r+2\,m}\frac{\partial}{\partial x_{p,q}}\left(\frac{C_{r+1\,s}}{D}\right)\right]u_{p,qm},
\end{aligned} \right\} \quad (10.40)
$$

(10.33) becomes

$$\delta^* l^{(i)} = T^{rs}\frac{\delta b_{rs}}{2} + (M^{rs} + N^{rs})\,\delta(u_{r,s}) + \lambda^{rs}\delta\mu_{rs} \tag{10.41}$$

which, according to (1.4), may be written

$$\delta^* l^{(i)} = [T^{ls}x^r_{,l} + M^{rs} + N^{rs}]\,\delta(u_{r,s}) + \lambda^{rs}\delta\mu_{rs} . \tag{10.42}$$

The form (10.42) of $\delta^* l^{(i)}$ is convenient for thermodynamic applications..

10*

§ 4. Use of the thermodynamic potential

1. The basic relation (2.23) still holds, provided $\delta * l^{(i)}$ is expressed by (10.42). Then the thermodynamic potential J is to be thought of as dependent on the actual state only through $x_{r,s}$, μ_{rs} and T, but, while in the symmetric case (2.23) permits expressions of Y_{rs}, T as derivatives of J, now that is not possible. In fact, it is easy to see that the identity

$$C_{rs}\mu^{rs} \equiv 0 \tag{10.43}$$

holds. Then $\delta(u_{p,q})$, $\delta\mu_{rs}$ are not independent, being connected by the relations

$$\sum_{rs}\left[\sum_{pq}\frac{\partial C_{rs}}{\partial x_{p,q}}(\delta u_{p,q})\,\mu_{rs} + C_{rs}\delta\mu_{rs}\right] = 0. \tag{10.44}$$

Therefore, equation (2.23) is not valid for arbitrary values of $\delta(u_{p,q})$, $\delta\mu_{rs}$ but only for those which satisfy (10.44). Then one deduces that

$$T_{ls}x_r^l + M_{rs} + N_{rs} + \tau\sum_{pq}\frac{\partial C_{pq}}{\partial x_{r,s}}\mu_{pq} = -\frac{\partial J}{\partial x_{r,s}}, \tag{10.45}$$

$$\lambda_{rs} + \tau C_{rs} = -\frac{\partial J}{\partial \mu_{r,s}}, \tag{10.46}$$

$$eE = -\frac{\partial J}{\partial T}, \tag{10.47}$$

where τ is a parameter which may depend even on the actual state, i.e., on $x_{r,s}$, μ_{rs}.

From (10.45) follows

$$T_{rs} = -\frac{1}{D}\sum_l C_{lr}\left[\frac{\partial J}{\partial x_{l,s}} + M_{ls} + N_{ls} + \tau\sum_{pq}\frac{\partial C_{pq}}{\partial x_{l,s}}\mu_{pq}\right]. \tag{10.48}$$

Knowledge of J and of the parameter τ determines T_{rs}, λ_{rs} and E, according to (10.46), (10.47), (10.48). In the isothermal case that is sufficient in order that the number of unknown functions in equations (10.25), (10.26), (10.28) be the same as the number of equations. The same thing happens in the adiabatic case, as it is possible to show by considerations analogous to the usual ones of the symmetric case.

2. The thermodynamic potential J and the parameter τ must satisfy the equations which express the symmetry conditions on T_{rs}. By (10.48), they are

$$\sum_l\left\{C_{lr}\left[\frac{\partial J}{\partial x_{l,s}} + M_{ls} + N_{ls} + \tau\sum_{pq}\frac{\partial C_{pq}}{\partial x_{l,s}}\mu_{pq}\right] - \right.$$
$$\left. - C_{ls}\left[\frac{\partial J}{\partial x_{l,r}} + M_{lr} + N_{lr} + \tau\sum_{pq}\frac{\partial C_{pq}}{\partial x_{l,r}}\mu_{pq}\right]\right\} = 0. \tag{10.49}$$

One may think that (10.49) determines the parameter τ, but it is not so, since (10.49) *only apparently* contains such a parameter. In fact, putting

$$
\left.
\begin{aligned}
\alpha_{rs} &= C_{lr} M'_{.s} - C_{ls} M'_{.r} , \\
\beta_{rs} &= C_{lr} N'_{.s} - C_{ls} N'_{.r} , \\
\gamma_{rs} &= \sum_{lpq} \left(C_{lr} \frac{\partial C_{rq}}{\partial x_{l,s}} - C_{ls} \frac{\partial C_{pq}}{\partial x_{l,r}} \right) \mu_{pq} ,
\end{aligned}
\right\}
\tag{10.50}
$$

it is possible to show that

$$
\left.
\begin{aligned}
\alpha_{s+1s} &= \frac{1}{2D} \sum_{lmpqtv} \left[C_{ls+1} \frac{\partial C_{lm}}{\partial x_{l,s}} - C_{ls} \frac{\partial C_{lm}}{\partial x_{l,s+1}} \right] u_{p,qm} \lambda_{vq} c_v \cdot c_p \times c_t , \\
\beta_{s+1s} &= \frac{1}{2D} \sum_{lq} [D u_{l,qs+2} - x_{l,s+2} \sum_{mt} C_{tm} u_{t,qm}] \lambda_{lq} , \\
\gamma_{s+1s} &= -\frac{1}{2D} \sum_{lmpqtv} \left[C_{ls+1} \frac{\partial C_{lm}}{\partial x_{l,s}} - C_{ls} \frac{\partial C_{lm}}{\partial x_{l,s+1}} \right] C_{vq} u_{p,qm} c_p \times c_v \cdot c_t
\end{aligned}
\right\}
$$

and that, if one supposes $\lambda_{rs} = -C_{rs}$,
$$\tag{10.51}$$

$$
\beta_{s+1s} = 0 , \qquad \alpha_{s+1s} + \gamma_{s+1s} = 0 \qquad [\text{for } \lambda_{rs} = -C_{rs}] , \tag{10.52}
$$

while it is plainly sufficient to consider (10.49) for $s = 1, 2, 3$ and $r = s + 1$.

Keeping in mind (10.40), one concludes that (10.49) is identically satisfied when one puts $\lambda_{rs} = -\tau C_{rs}$, $\frac{\partial J}{\partial x_{p,q}} = 0$. This means that in (10.49) the coefficient of τ is identically equal to zero, in view of (10.40). By what is said above, it is evident that (10.49) is equivalent to the equation

$$
\alpha_{s+1s} + \beta_{s+1s} + \sum_l \left[C_{ls+1} \frac{\partial J}{\partial x_{l,s}} - C_{ls} \frac{\partial J}{\partial x_{l,s+1}} \right] = 0 , \tag{10.53}
$$

which is to be considered supposing $\lambda_{rs} = -\frac{\partial J}{\partial \mu_{rs}}$ in the expressions (10.51) of $\alpha_{s+1s}, \beta_{s+1s}$.

Plainly [see (10.51)] the coefficients of equation (10.53) depend on the second derivatives of the displacement, while J depends on them only through μ_{rs}, but it is possible to show that, substantially, the coefficients of (10.53) depend on $u_{r,st}$ only through μ_{pq}. Therefore (10.53) is a set of three differential equations which restrict the analytical constitution of the thermodynamic potential.

I observe that (10.51, 1) may be written in the form

$$
\alpha_{s+1s} = \frac{1}{2D} \sum_{lmq} [u_{l+1,qm} (C_{l+1m} x_{l,s+2} - C_{lm} x_{l+1,s+2}) +
$$

$$
+ u_{l+2,qm} (C_{l+2m} x_{l,s+2} - C_{lm} x_{l+2,s+2})] \lambda_{lq} , \tag{10.54}
$$

while (10.53), by virtue of (10.51, 2), (10.54) becomes

$$\sum_l \left\{ C_{ls+1} \frac{\partial J}{\partial x_{l,s}} - C_{ls} \frac{\partial J}{\partial x_{l,s+1}} - \right.$$

$$\left. - \frac{1}{2D} \sum_q \left[D u_{l,qs+2} - \sum_{mt} C_{lm} x_{l,s+2} u_{t,qm} \right] \frac{\partial J}{\partial \mu_{lq}} \right\} = 0. \tag{10.55}$$

In view of the equality

$$\left. \begin{array}{l} C_{\nu+1s+1} C_{\nu s+2} - C_{\nu+1s+2} C_{\nu s+1} = - x_{\nu+2,s} D , \\ C_{\nu+2s+1} C_{\nu s+2} - C_{\nu+2s+2} C_{\nu s+1} = x_{\nu+1,s} D , \end{array} \right\} \tag{10.56}$$

it is easy to recognize that

$$\sum_{ls} C_{\nu s+2} \left[C_{ls+1} \frac{\partial J}{\partial x_{l,s}} - C_{ls} \frac{\partial J}{\partial x_{l,s+1}} \right] = D \sum_s \left[\frac{\partial J}{\partial x_{\nu+2,s}} x_{\nu+1,s} \right.$$

$$\left. - \frac{\partial J}{\partial x_{\nu+1,s}} x_{\nu+2,s} \right] \tag{10.57}$$

and, further, according to (10.39), that

$$\sum_{lqs} C_{\nu s+2} \left[D u_{l,qs+2} - \sum_{mt} C_{lm} x_{l,s+2} u_{t,qm} \right] \frac{\partial J}{\partial \mu_{lq}}$$

$$= D \sum_{lqm} (C_{\nu m} u_{l,qm} - C_{lm} u_{\nu,qm}) \frac{\partial J}{\partial \mu_{lq}}$$

$$= - 2 D^2 \sum_q \left[\frac{\partial J}{\partial \mu_{\nu+2q}} \mu_{\nu+1q} - \frac{\partial J}{\partial \mu_{\nu+1q}} \mu_{\nu+2q} \right]. \tag{10.58}$$

Since the determinant of C_{rs} is not zero, the set of equations (10.55) is equivalent to the one deduced by multiplying the equation of (10.55) by $C_{\nu s+2}$, summing with respect to s and varying ν from 1 to 3. Specifically, keeping in mind (10.57), (10.58), the set of equations (10.55) is modified, and becomes

$$\sum_q \left\{ \frac{\partial J}{\partial x_{\nu+2,q}} x_{\nu+1,q} - \frac{\partial J}{\partial x_{\nu+1,q}} x_{\nu+2,q} + \frac{\partial J}{\partial \mu_{\nu+2q}} \mu_{\nu+1q} - \frac{\partial J}{\partial \mu_{\nu+1q}} \mu_{\nu+2q} \right\} = 0, \tag{10.59}$$

where the coefficients depend only on x_{rs}, μ_{rs} and represent the symmetry conditions (10.49) in the most suitable form. Naturally, J must satisfy the *fundamental condition* of elastic bodies [see Chapter III, § 2]. Then if $J \equiv 0$ in C^*, we must have

$$\int_{C^*} J dC^* > 0, \tag{10.60}$$

corresponding to any isothermal non-rigid displacement starting from $C*$. For homogeneous bodies, on the assumption that $J \equiv 0$ in $C*$, from (10.60) follows $J > 0$. It is easy to show that (10.59) is satisfied if J depends on $x_{p,q}$ through b_{rs} [see (1.4)] and on μ_{pq} only by means of ν_{rs}, with

$$\nu_{rs} = \mu_{ir}\,\mu^i_{.s}\,. \tag{10.61}$$

§ 5. Equilibrium equations in the linear theory

1. Let us suppose that the external forces and couples are proportional to a parameter θ and that for u_r, Y_{rs}, etc., any number of differentiations with respect to θ are permitted, as was assumed in Chapters IV, V, for the solutions in the symmetric case.

Plainly

$$\mu^{(0)}_{rs} = 0\,, \qquad \mu^{(1)}_{rs} = \omega^{(1)}_{r,s}\,, \tag{10.62}$$

with

$$\omega^{(1)}_r = \frac{1}{2}\,(u^{(1)}_{r+2,\,r+1} - u^{(1)}_{r+1,\,r+2})\,. \tag{10.63}$$

By a procedure analogous to the one used in Chapter IV one shows that the stress in $C*$, if not equal to zero, satisfies the equations

$$\left.\begin{aligned} Y^{(0),\,m}_{rm} &= 0 \qquad\qquad (\text{in } C*)\,, \\ Y^{(0)}_{r+1\,r+2} - Y^{(0)}_{r+2\,r+1} &= \lambda^{(0),\,m}_{rm}\,, \end{aligned}\right\} \tag{10.64}$$

$$\left.\begin{aligned} Y^{(0)}_{rm}\,N^m_* &= 0 \\ \lambda^{(0)}_{rm}\,N^m_* &= 0\,. \end{aligned}\quad (\text{on } \Sigma*)\,,\right\} \tag{10.65}$$

Thus it is evident that the stress is necessarily symmetric in a theory for which the surface moments are equal to zero.

2. The set of equations corresponding to the first derivative of u_r with respect to θ, for $\theta = 0$, is

$$\left.\begin{aligned} Y^{(1),\,m}_{rm} &= k* F_r - (Y^{(0)\,lm}\,u^{(1)}_{r,l})_{,m} \qquad\qquad (\text{in } C*)\,, \\ \lambda^{(1),\,m}_{rm} &= Y^{(1)}_{r+1\,r+2} - Y^{(1)}_{r+2\,r+1} - u^{(1)}_{r+1,l}(Y^{(0)\cdot l}_{.r+2} - Y^{(0)l}_{.r+2}) - \\ &\quad - u^{(1)}_{r+2,l}(Y^{(0)l}_{.r+1} - Y^{(0)\cdot l}_{r+1}) + k* M_r \quad (\text{in } C*)\,, \end{aligned}\right\} \tag{10.66}$$

$$\left.\begin{aligned} Y^{(1)}_{rm}\,N^m_* &= f^*_r - Y^{(0)}_{lm}\,u^{(1),\,l}_r\,N^m_* \\ \lambda^{(1)}_{rm}\,N^m_* &= m^*_r\,. \end{aligned}\quad (\text{on } \Sigma*)\,,\right\} \tag{10.67}$$

Since

$$T^{(i)}_{rs} = \frac{1}{2}\,(Y^{(i)}_{rs} + Y^{(i)}_{sr})\,, \tag{10.68}$$

consequently (10.64, 1), (10.65, 1) become

$$T^{(0),m}_{rm} + \frac{1}{2} [\lambda^{(0)}_{r+2s,r+1} - \lambda^{(0)}_{r+1s,r+2}]^{\cdot s} = 0 \qquad \text{(in } C^*\text{)},$$

$$T^{(0)}_{rm} N^m_* + \frac{1}{2} [\lambda^{(0),s}_{r+2s} N^*_{r+1} - \lambda^{(0),s}_{r+1s} N^*_{r+2}] = 0 \qquad \text{(on } \Sigma^*\text{)}.$$

$$(10.69)$$

Analogously, (10.66, 1), (10.67, 1) become

$$T^{(1),m}_{rm} + \frac{1}{2} [\lambda^{(1)}_{r+2s,r+1} - \lambda^{(1)}_{r+1s,r+2}]^{\cdot s} = - \left[u^{(1)}_{r,l} T^l_{(0)m} - \right.$$

$$- \frac{1}{2} u^{(1)}_{m,l} (Y^{rl}_{(0)} - Y^{lr}_{(0)}) \Big]^{\cdot m} + k^* F_r + \frac{1}{2} [(k^* M_{r+2}),_{r+1} -$$

$$- (k^* M_{r+1}),_{r+2}] \qquad \text{(in } C^*\text{)},$$

$$T^{(1)}_{rm} N^m_* + \frac{1}{2} [\lambda^{\cdot s}_{r+2s} N^*_{r+1} - \lambda^{\cdot s}_{r+1s} N^*_{r+2}] = - \left[u^{(1)}_{r,l} T^{(0)l}_m - \right.$$

$$- \frac{1}{2} u^{(1)}_{m,l} (Y^{rl}_{(0)} - Y^{lr}_{(0)}) \Big] N^m_* + f^*_r + \frac{k^*}{2} (M_{r+2} N^*_{r+1} -$$

$$- M_{r+1} N^*_{r+2}) \qquad \text{(on } \Sigma^*\text{)}.$$

$$(10.70)$$

Equations (10.70) together with (10.66, 2), (10.67, 2) represent the basic set of differential equations of the linear theory, valid for slightly deformable bodies. If C^* is a natural equilibrium state [unstressed state], equations (10.70), (10.66, 2) become

$$T^{(1),m}_{rm} + \frac{1}{2} [\lambda^{(1)}_{r+2s,r+1} - \lambda^{(1)}_{r+1s,r+2}]^{\cdot s} = k^* F_r +$$

$$+ \frac{1}{2} [(k^* M_{r+2}),_{r+1} - (k^* M_{r+1}),_{r+2}] \qquad \text{(in } C^*\text{)},$$

$$T^{(1)}_{rm} N^m_* + \frac{1}{2} [\lambda^{(1),s}_{r+2s} N^*_{r+1} - \lambda^{(1),s}_{r+1s} N^*_{r+2}] = f^*_r +$$

$$+ \frac{k^*}{2} [M_{r+2} N^*_{r+1} - M_{r+1} N^*_{r+2}] \qquad \text{(on } \Sigma^*\text{)},$$

$$(10.71)$$

$$\lambda^{(1),m}_{rm} = Y^{(1)}_{r+1\,r+2} - Y^{(1)}_{r+2\,r+1} + k^* M_r,$$

$$\lambda^{(1)}_{rm} N^m_* = m^*_r, \qquad \text{(on } \Sigma)^*.$$

$$(10.72)$$

The equality

$$Y^{(1)}_{rs} + Y^{(1)}_{sr} = 2 T^{(1)}_{rs} \qquad (10.73)$$

is to be associated with (10.71), (10.72).

§ 6. Isothermal elastic potential for the small deformations of isotropic bodies

Let us suppose that C^* is a natural equilibrium configuration. Under the hypothesis of small deformations it is natural to assume the validity of linear stress-strain relations [Hooke's Law].

This result is reached by differentiation with respect to θ, following a procedure parallel to that of Chapter IV. Further, the deduction of the stress-strain relations in the linear case, starting from the theory of finite displacements, is profitable, for only in such manner is it easy to remove the indeterminancy of the parameter τ.

From (10.46) follows

$$\lambda_{rs}^{(0)} = -\tau^{(0)} \delta_{rs} - \left(\frac{\partial J}{\partial \mu_{rs}} \right)^{(0)}. \tag{10.74}$$

Then, C^* being an unstressed state, we must have

$$\tau^{(0)} + \left(\frac{\partial J}{\partial \mu_{rr}} \right)^{(0)} = 0, \qquad \left(\frac{\partial J}{\partial \mu_{rs}} \right)^{(0)} = 0, \qquad \text{for } r \neq s. \tag{10.75}$$

From (10.48) follows

$$T_{rs}^{(0)} = -\left(\frac{\partial J}{\partial x_{r,s}} \right)^{(0)} = 0. \tag{10.76}$$

From (10.46), (10.48) one deduces

$$T_{rs}^{(1)} = -\tau^{(0)} \sum_{pq} \left(\frac{\partial C_{rq}}{\partial x_{r,s}} \right)^{(0)} \mu_{pq}^{(1)} - \sum_{pq} \left[\left(\frac{\partial^2 J}{\partial x_{r,s} \partial x_{p,q}} \right)^{(0)} u_{p,q}^{(1)} + \right.$$
$$\left. + \left(\frac{\partial^2 J}{\partial x_{r,s} \partial \mu_{p,q}} \right)^{(0)} \mu_{pq}^{(1)} \right], \tag{10.77}$$

$$\lambda_{rs}^{(1)} = -\tau^{(0)} C_{rs}^{(1)} - \tau^{(1)} C_{rs}^{(0)} - \sum_{pq} \left[\left(\frac{\partial^2 J}{\partial x_{r,q} \partial \mu_{rs}} \right)^{(0)} u_{p,q}^{(1)} + \left(\frac{\partial^2 J}{\partial \mu_{rs} \partial \mu_{pq}} \right)^{(0)} \mu_{pq}^{(1)} \right]. \tag{10.78}$$

It is easy to recognize that

$$\left. \begin{array}{l} C_{rs}^{(0)} = \delta_{rs}, \qquad C_{rs}^{(1)} = \delta_{rs} (u_{r+1,\,r+1}^{(1)} + u_{r+2,\,r+2}^{(1)}) - \delta_{rs+1} u_{r+2,\,r}^{(1)} - \delta_{rs+2} u_{r+1,\,r}^{(1)}, \\ \displaystyle\sum_{pq} \left(\frac{\partial C_{pq}}{\partial x_{r,s}} \right)^{(0)} \mu_{pq}^{(1)} = -\delta_{rs} \mu_{rr}^{(1)} - \delta_{rs+1} \mu_{r+2r}^{(1)} - \delta_{rs+2} \mu_{r+1r}^{(1)}. \end{array} \right\} \tag{10.79}$$

For $r \neq s$, from (10.77), (10.78), (10.79) follows

$$T_{rs}^{(1)} = \tau^{(0)} (\delta_{rs+1} \mu_{r+2r}^{(1)} + \delta_{rs+2} \mu_{r+1r}^{(1)}) - \sum_{pq} \left[\left(\frac{\partial^2 J}{\partial x_{r,s} \partial x_{p,q}} \right)^{(0)} u_{p,q}^{(1)} + \right.$$
$$\left. + \left(\frac{\partial^2 J}{\partial x_{r,s} \partial \mu_{pq}} \right)^{(0)} \mu_{pq}^{(1)} \right], \qquad r \neq s \tag{10.80}$$

$$\lambda_{rs}^{(1)} = \tau^{(0)} (\delta_{rs+1} u_{r+2r}^{(1)} + \delta_{rs+2} u_{r+1r}^{(1)}) - \sum_{pq} \left[\left(\frac{\partial^2 J}{\partial x_{p,q} \partial \mu_{rs}} \right)^{(0)} u_{p,q}^{(1)} + \right.$$
$$\left. + \left(\frac{\partial^2 J}{\partial \mu_{rs} \partial \mu_{pq}} \right)^{(0)} \mu_{pq}^{(1)} \right]. \tag{10.81}$$

From (10.59), by differentiation with respect to θ, for $\theta = 0$, and (10.75), one shows that

$$\sum_{lm}\left\{\left[\left(\frac{\partial^2 J}{\partial x_{\nu+2,\,\nu+1}\,\partial x_{l,m}}\right)^{(0)}-\left(\frac{\partial^2 J}{\partial x_{\nu+1,\,\nu+2}\,\partial x_{l,m}}\right)^{(0)}\right]u^{(1)}_{l,m}+\left[\left(\frac{\partial^2 J}{\partial x_{\nu+2,\,\nu+1}\,\partial \mu_{lm}}\right)^{(0)}-\right.\right.$$
$$\left.\left.-\left(\frac{\partial^2 J}{\partial x_{\nu+1,\,\nu+2}\,\partial \mu_{lm}}\right)^{(0)}\right]\mu^{(1)}_{lm}\right\}+\tau^{(0)}(\mu^{(1)}_{\nu+2\,\nu+1}-\mu^{(1)}_{\nu+1\,\nu+2})=0. \qquad (10.82)$$

I shall presume that for any infinitesimal irrotational displacement starting from C^* [for which $\mu^{(1)}_{rs}=0$] the quantities $\Psi^{(1)}_{rs}$, $\lambda^{(1)}_{rs}$ are all equal to zero. Therefore, when

$$u^{(1)}_{r,r+1}-u^{(1)}_{r+1,r}=0 \qquad (10.83)$$

and, consequently, when $\mu^{(1)}_{rs}=0$, the $\lambda^{(1)}_{rs}$ must all be equal to zero. From (10.81), for $s=r+1$, one shows that in order for this to happen, we must have

$$\tau^{(0)}-\left(\frac{\partial^2 J}{\partial x_{r+1,r}\,\partial \mu_{rr+1}}\right)^{(0)}-\left(\frac{\partial^2 J}{\partial x_{r,r+1}\,\partial \mu_{rr+1}}\right)^{(0)}=0, \qquad (10.84)$$

while from (10.82) follows

$$\tau^{(0)}+\left(\frac{\partial^2 J}{\partial x_{r,r+1}\,\partial \mu_{rr+1}}\right)^{(0)}-\left(\frac{\partial^2 J}{\partial x_{r+1,r}\,\partial \mu_{r\,r+1}}\right)^{(0)}=0. \qquad (10.85)$$

From (10.84), (10.85) follows

$$\left(\frac{\partial^2 J}{\partial x_{r,r+1}\,\partial \mu_{rr+1}}\right)^{(0)}=0, \qquad \tau^{(0)}-\left(\frac{\partial^2 J}{\partial x_{r+1,r}\,\partial \mu_{rr+1}}\right)^{(0)}=0. \qquad (10.86)$$

As I shall show, these equalities are sufficient to prove that in the case of isotropic slightly deformable bodies, which are the only ones I shall consider, the parameter τ may be supposed zero, while J is equal to the sum of a function J_1 of the $x_{r,s}$ only and of a function J_2 of the μ_{rs} only. Actually (10.86, 1) shows that the expression of $T^{(1)}_{r\,r+1}$ deduced from (10.80) for $s=r+1$ does not contain a term in μ_{rr+1}. On the other hand a linear homogeneous expression for $T^{(1)}_{rs}$ corresponding to the isothermal case,

$$T^{(1)}_{rs}=m_{rspq}\,u^{(1)\,p,q}+n_{rspq}\,\mu^{(1)\,pq}, \qquad (10.87)$$

corresponds to the isotropic case if, and only if, the tensors m_{rspq}, n_{rspq} are isotropic; in a Cartesian reference frame, then,

$$\left.\begin{aligned}m_{rspq}&=\alpha\delta_{rs}\delta_{pq}+\beta\delta_{rp}\delta_{sq}+\gamma\delta_{rq}\delta_{sp},\\ n_{rspq}&=\alpha'\delta_{rs}\delta_{pq}+\beta'\delta_{rp}\delta_{sq}+\gamma'\delta_{rq}\delta_{sp},\end{aligned}\right\} \qquad (10.88)$$

where α, β, etc., are arbitrary coefficients. Then it follows that

$$T^{(1)}_{rs}=\alpha\delta_{rs}u^{(1),\,p}_p+\beta u^{(1)}_{r,s}+\gamma u^{(1)}_{s,r}+\alpha'\delta_{rs}\mu^{(1),\,p}_p+\beta'\mu_{rs}+\gamma'\mu_{sr}, \qquad (10.89)$$

which, by the symmetry of $T_{rs}^{(1)}$, implies

$$\beta = \gamma , \qquad \beta' = \gamma' . \tag{10.90}$$

Therefore,

$$T_{rs}^{(1)} = \alpha \delta_{rs} \varepsilon_p^{(1)p} + 2\beta \varepsilon_{rs}^{(1)} + \beta' (\mu_{rs} + \mu_{sr}) , \tag{10.91}$$

and since the expression of $T_{rr+1}^{(1)}$ *must not contain* a term in $\mu_{rr+1}^{(1)}$, we must have

$$\beta' = 0 . \tag{10.92}$$

Thus the *expression for $T_{rs}^{(1)}$ is analogous to that of the symmetric case.*

Analogous considerations show that a linear homogeneous expression for $\lambda_{rs}^{(1)}$ in the isotropic case is of the type

$$\lambda_{rs}^{(1)} = A' \delta_{rs} \varepsilon_p^{(1)p} + B' u_{r,s}^{(1)} + C' u_{s,r}^{(1)} - (B\mu_{rs} + C\mu_{sr}) , \tag{10.93}$$

which, since it must vanish for any irrotational displacement, implies that

$$A' = 0 , \qquad B' + C' = 0 , \tag{10.94}$$

while (10.81), written for $s = r + 1$, according to (10.86, 1) shows that *the expression for $\lambda_{rr+1}^{(1)}$ cannot contain a term in $u_{r,r+1}^{(1)}$.* Therefore we must have

$$B' = C' = 0 . \tag{10.95}$$

It is evident, by (10.91), (10.92), (10.93), (10.94), (10.95), that the expression for $T_{rs}^{(1)}$ depends only on $\varepsilon_{pq}^{(1)}$, while that of $\lambda_{rs}^{(1)}$ depends only on $\mu_{pq}^{(1)}$.

Upon introduction of the quadratic form J_2, which represents the second derivative of J with respect to θ for $\theta = 0$, considerations analogous to those in Chapter IV show that $T_{rs}^{(1)}, \lambda_{rs}^{(1)}$ coincide with the derivatives of J_2. Therefore, denoting by W the corresponding isothermal elastic potential for isothermal infinitesimal displacements, one may suppose that

$$W = W' + W'' , \tag{10.96}$$

where W' depends only on $\varepsilon_{rs}^{(1)}$ and W'' only on $\mu_{rs}^{(1)}$.

From (10.93), (10.94), (10.95) follows

$$W'' = \frac{1}{2} (B\mu_{pq} + C\mu_{qp}) \mu^{pq} . \tag{10.97}$$

Taking the second derivative of (10.59) with respect to θ for $\theta = 0$, and keeping in mind (10.75), one deduces

$$\sum_q \left[\left(\frac{\partial J}{\partial \mu_{\nu+1q}} \right)^{(1)} \mu_{\nu+2q}^{(1)} - \left(\frac{\partial J}{\partial \mu_{\nu+2q}} \right)^{(1)} \mu_{\nu+1q}^{(1)} \right] = 0 , \tag{10.98}$$

equivalent to

$$\sum_{qlm} \left[\left(\frac{\partial^2 J_2}{\partial \mu_{\nu+1q} \partial \mu_{lm}} \right)^{(0)} \mu_{\nu+2q}^{(1)} - \left(\frac{\partial^2 J_2}{\partial \mu_{\nu+2q} \partial \mu_{lm}} \right)^{(0)} \mu_{\nu+1q}^{(1)} \right] \mu_{lm}^{(1)} = 0, \quad (10.99)$$

which, according to (10.97), becomes

$$C \left[\mu_{\nu+2q}^{(1)} \mu_{\cdot \nu+1}^{(1)q} - \mu_{\nu+1q}^{(1)} \mu_{\cdot \nu+2}^{(1)q} \right] = 0 \qquad (10.100)$$

and implies

$$C = 0. \qquad (10.101)$$

Naturally, from the condition $J_2 > 0$ [see § 4] it follows that the quadratic form $W' + W''$ must be positive-definite. Then from (10.96), (10.97), (10.101) follows

$$B \geqslant 0, \qquad W' > 0. \qquad (10.102)$$

Regarding the physical meaning of the coefficient B, I observe only that considerations about the propagation of discontinuity waves in a linear isotropic elastic medium show that it is connected with the size of the elements of which the medium is to be thought constituted, so that the asymmetric theory may be acceptable[1].

§ 7. Recapitulation of the general equations valid in the case of infinitesimal isothermal transformations

From what has been said in the two previous sections it follows that the basic equations valid in the statics of slightly deformable bodies with asymmetric components of stress, supposing C^* the natural state and suppressing the index (1), are

$$Y_{rs} + Y_{sr} = 2 T_{rs}, \qquad (10.103)$$

$$Y_{r+1\,r+2} - Y_{r+2\,r+1} = \lambda_{rm}^m - k^* M_r, \qquad (10.104)$$

$$\lambda_{rm} N_*^m = m_r^* \qquad (\text{on } \Sigma^*), \qquad (10.105)$$

$$\left.\begin{array}{l} T_{rm}^m + \dfrac{1}{2} \left(\lambda_{r+2s,\,r+1} - \lambda_{r+1s,\,r+2} \right)^{\cdot s} = k^* F_r + \\[2mm] \qquad + \dfrac{1}{2} \left[(k^* M_{r+2})_{,r+1} - (k^* M_{r+1})_{,r+2} \right] \qquad (\text{in } C^*), \\[3mm] T_{rm} N_*^m + \dfrac{1}{2} \left(\lambda_{r+2s}^{\cdot s} N_{r+1}^* - \lambda_{r+1s}^{\cdot s} N_{r+2}^* \right) = f_r^* + \\[2mm] \qquad + \dfrac{k^*}{2} \left(M_{r+2} N_{r+1}^* - M_{r+1} N_{r+2}^* \right) \qquad (\text{on } \Sigma^*). \end{array}\right\} \quad (10.106)$$

---

[1] For example, if it is supposed that the elements are small cubes whose edges are equal to l, one finds $l = 2 \sqrt{\dfrac{3B}{\mu}}$, where μ denotes that of the two coefficients of Lamé which is always positive [GRIOLI, 12].

To these equations the following are to be associated, valid in the case of isotropic bodies and for isothermal displacements:

$$W = W'(\varepsilon) + W''(\mu) , \tag{10.107}$$

$$W' = \frac{1}{2} \left[(\lambda + 2\mu)(\varepsilon_r^{\cdot r})^2 - 4\mu \sum_r (\varepsilon_{rr}\varepsilon_{r+1\,r+1} - \varepsilon_{rr+1}^2) \right], \tag{10.108}$$

$$W'' = \frac{B}{2} \mu_{rs} \mu^{rs} , \qquad B \geqslant 0 , \tag{10.109}$$

$$\varepsilon_{rs} = \frac{1}{2}(u_{r,s} + u_{s,r}) , \qquad \mu_{rs} = \frac{1}{2}(u_{r+2,\,r+1} - u_{r+1,\,r+2})_{,s} , \tag{10.110}$$

$$T_{rs} = -\frac{\partial W}{(2 - \delta_{rs})\,\partial \varepsilon_{rs}} , \qquad \lambda_{rs} = -\frac{\partial W}{\partial \mu_{rs}} , \tag{10.111}$$

$$\delta^* l^{(i)} = T^{rs} \delta \varepsilon_{rs} + \lambda^{rs} \delta \mu_{rs} . \tag{10.112}$$

(10.104) shows that in a theory in which surface moments λ_{rs} may be non-zero, the stress components generally are asymmetric, even if the body moments and the external surface moments are zero [$M_r = 0$, $m_r^* = 0$]. This fact allows one to obtain regular solutions where the symmetric theory would give singular solutions.

Instead, in a theory where λ_{rs}, m_r^* are supposed necessarily zero [$B = 0$], equations (10, 104), (10.106) become[1]

$$Y_{r+1\,r+2} - Y_{r+2\,r+1} = -k^* M_r , \tag{10.113}$$

$$\left. \begin{array}{l} T_{rm}^{\cdot m} = k^* F_r + \dfrac{1}{2} \left[(k^* M_{r+2})_{,r+1} - (k^* M_{r+1})_{,r+2} \right] \quad \text{(in } C^*) \\[2ex] T_{rm} N_*^m = f_r^* + \dfrac{k^*}{2} \left[M_{r+2} N_{r+1}^* - M_{r+1} N_{r+2}^* \right] \qquad \text{(on } \Sigma^*) \end{array} \right\} \tag{10.114}$$

while equations (10.105) are not to be considered any longer. Further, (10.112) becomes

$$\delta^* l^{(i)} = T^{rs} \delta \varepsilon_{rs} . \tag{10.115}$$

Since $\delta^* l^{(i)} = -\delta W$ for any isothermal transformation, it follows that

$$\delta W = -T^{rs} \delta \varepsilon_{rs} . \tag{10.116}$$

Expressions (10.115), (10.116) are formally identical to those of the symmetric case. Then *the analytical form of W does not change in passing from the symmetric to the asymmetric case if the existence of surface moments is not assumed*[2].

---

[1] Contrary to (10.104), (10.113), according to SOMIGLIANA, BODASZEWSKI, the differences $Y_{rs} - Y_{sr}$ depend on the local rotations.

[2] Equation (10.116) is very different from the corresponding one in SOMIGLIANA's paper.

(10.114) shows that the integral properties valid for Y_{rs} [Chapter VI], are also valid for T_{rs}, if the definitions of the *astatic, hyper-astatic* and *n-astatic* coordinates are suitably modified. Analogous integral properties certainly may be established even if $\lambda_{rs} \neq 0$. Further, the theorems of Betti, Castigliano, Menabrea, etc., *must* hold. Then it *must* be possible to establish an integration method analogous to the one of Chapter VII.

§ 8. Discussion of an example

Let C^* be a body having the form of a right prism of square cross-section of side a, without body forces and subjected to external surface forces acting only upon the lateral surface. Let the rectangular Cartesian reference frame have its origin in the mid-point of a lateral edge, the axis of y_3 parallel to it and the axes of y_1, y_2 parallel to the sides of the cross-section.

Supposing that vector f^*, characterizing the external surface forces, is perpendicular to axis y_3 and independent of this coordinate, solutions are to be found for which

$$u_3 \equiv 0, \qquad u_1, u_2 \text{ independent of } y_3. \tag{10.117}$$

I shall denote by f_{is} the component of index i of f^* on the side $y_s = 0$ and by $f_{is}^{(a)}$ the analogous one on the side $y_s = a$ ($s = 1, 2$). According to (10.107), (10.108), (10.109), (10.110) equations (10.106, 1) reduce to only two, namely,

$$\mu \Delta_2 u_r + (\lambda + \mu) \sum_{s=1}^{2} u_{s,rs} + \frac{B}{4} \Delta_2 \left[\sum_{s=1}^{2} u_{s,rs} - \Delta_2 u_r \right] = 0 \qquad (r = 1, 2), \tag{10.118}$$

while the associated boundary equations [see (10.105), (10.106, 2)] are the following:

$$(u_{2,1} - u_{1,2})_{,1} = 0, \qquad \text{for } y_1 = 0, a, \tag{10.119}$$

$$(u_{2,1} - u_{1,2})_{,2} = 0, \qquad \text{for } y_2 = 0, a, \tag{10.120}$$

$$(\lambda + 2\mu) u_{1,1} + \lambda u_{2,2} = \begin{cases} -f_{11}, & \text{for } y_1 = 0, \\ f_{11}^{(a)}, & \text{for } y_1 = a, \end{cases} \tag{10.121}$$

$$(\lambda + 2\mu) u_{2,2} + \lambda u_{1,1} = \begin{cases} -f_{22}, & \text{for } y_2 = 0, \\ f_{22}^{(a)}, & \text{for } y_2 = a, \end{cases} \tag{10.122}$$

$$\mu (u_{2,1} + u_{1,2}) - \frac{B}{4} \Delta_2 (u_{2,1} - u_{1,2}) = \begin{cases} -f_{21} & \text{for } y_1 = 0, \\ f_{21}^{(a)} & \text{for } y_1 = a, \end{cases} \tag{10.123}$$

$$\mu (u_{2,1} + u_{1,2}) + \frac{B}{4} \Delta_2 (u_{2,1} - u_{1,2}) = \begin{cases} -f_{12} & \text{for } y_2 = 0, \\ f_{12}^{(a)} & \text{for } y_2 = a. \end{cases} \tag{10.124}$$

Let us suppose

$$f_{11} = 0, \qquad f_{12} = 0. \qquad (10.125)$$

Then one has shearing forces upon the side $y_1 = 0$ and normal forces upon the side $y_2 = 0$.

It is easy to verify that the functions

$$u_1 = b \left\{ y_1^2 - \frac{\lambda + 2\mu}{\mu} y_2^2 - B \frac{\lambda + 2\mu}{2\mu^2} \frac{\operatorname{sh}\beta y_2 + \operatorname{sh}\beta(a - y_2)}{\operatorname{sh}\beta a} \right\},$$
$$u_2 = 0,$$
$$(10.126)$$

where b is a constant, and

$$\beta = \sqrt{\frac{4\mu}{B}}, \qquad (10.127)$$

satisfy equations (10.118) and, according to (10.125), (10.127), satisfy the boundary conditions (10.119), (10.120), (10.121), (10.124, 1). Further, according to (10.123, 1), (10.127), on the side $y_1 = 0$ we have

$$f_{21} = 2(\lambda + 2\mu) b \left[y_2 + \sqrt{\frac{B}{\mu}} \frac{\operatorname{ch}\beta y_2 - \operatorname{ch}\beta(a - y_2)}{\operatorname{sh}\beta a} \right]. \qquad (10.128)$$

Then

$$\lim_{y_2 = 0} f_{21} = 2(\lambda + 2\mu) b \sqrt{\frac{B}{\mu}} \frac{1 - \operatorname{ch}\beta a}{\operatorname{sh}\beta a}. \qquad (10.129)$$

In other words, on the side $y_1 = 0$ one has a shearing force which, for $y_2 \to 0$, tends to a limit different from zero, while on side $y_2 = 0$ the external forces are normal. Therefore, we have

$$\lim_{y_2 \to 0} Y_{21}(0, y_2) = 2(\lambda + 2\mu) b \sqrt{\frac{B}{\mu}} \frac{1 - \operatorname{ch}\beta a}{\operatorname{sh}\beta a} \neq 0,$$
$$\lim_{y_1 \to 0} Y_{12}(y_1, 0) = 0.$$
$$(10.130)$$

In general, one has

$$Y_{21} - Y_{12} = -\lambda_{3s}^{\cdot s} = -\frac{B}{2} u_{1,222} = -2b(\lambda + 2\mu) \sqrt{\frac{B}{\mu}} \frac{\operatorname{ch}\beta y_2 - \operatorname{ch}\beta(a - y_2)}{\operatorname{sh}\beta a}.$$
$$(10.131)$$

Therefore, in a problem where the external shearing forces acting upon the side $y_1 = 0$, $y_2 = 0$ have different limits when one tends toward the lateral edge, the introduction of λ_{rs} makes the solution regular and, in particular, gives single-valued stress. Instead, the symmetric theory implies a singularity on the lateral edge which is not due to the corner but to the nature of the external forces. In fact, it is easy to see that if B tends to zero, (10.126) tends to

$$u_1 = b \left(y_1^2 - \frac{\lambda + 2\mu}{\mu} y_2^2 \right), \qquad u_2 = 0, \qquad (10.132)$$

which represents the solution of the corresponding problem studied according to the symmetric theory. This solution is regular together with its derivatives but corresponds to external shearing forces which tend to the same limit when one approaches the prism's lateral edge $y_1 = y_2 = 0$.

In fact, we have

$$\left. \begin{array}{l} \lim_{B \to 0} (Y_{21} - Y_{12}) = 0, \\[2mm] \lim_{B \to 0} f_{21} = 2b (\lambda + 2\mu) y_2, \\[2mm] \lim_{y_1 \to 0} \lim_{B \to 0} f_{21} = 0. \end{array} \right\} \qquad (10.133)$$

Instead, if one studies by the symmetric theory $[B = 0]$ a problem with external shearing forces having different limits when one approaches the lateral edge, one finds multi-valued stress with divergent derivatives. For example, this happens (REISSNER) if the external shearing forces are zero on the side $y_2 = 0$ and equal to an arbitrary constant on the side $y_1 = 0$.

By the above considerations one sees that a theory where surface moments are present may often be appropriate.

Bibliography

ADKINS, J. E.: Some general results in the theory of large elastic deformations. Proc. roy. Soc. **A**, **231**, 75—90 (1955).

AMERIO, L.: Sul calcolo delle soluzioni dei problemi al contorno per le equazioni lineari del secondo ordine di tipo ellittico. Am. J. Math. LXIX, **3**, 447 (1947).

BAKER, M., and J. L. ERICKSEN: Inequalities restricting the form of the stress-deformation relations for isotropic elastic solids and Reiner-Rivlin fluids. J. Washington Acad. Sci. **44**, 33—35 (1954).

BERNSTEIN, B., and J. ERICKSEN: Work functions in hypo-elasticity. Arch. rational Mech. Anal. **1**, 396—409 (1958).

BLACKBURN, W. S., and A. E. GREEN: Second-order torsion and bending of isotropic elastic cylinders. Proc. roy. Soc. **A**, **240**, 408—422 (1957).

BODASZEWSKI, S.: On the asymmetric state of stress and its applications to the mechanics of continuous medium. Arch. Med. Stos. **5**, 351—396 (1953).

BORDONI, P. G.: [1] Sopra le trasformazioni termoelastiche finite di certi solidi omogenei ed isotropi. Rend. Mat. e Appl. Vol. XII S. V 237—266 (1953). — [2] Trasformazioni adiabatiche di ampiezza finita. Ric. scient. A 23, **9**, 1569—1578 (1953). — [3] Deduzione di un'equazione di stato dei solidi dalla teoria delle trasformazioni termoelastiche finite. Rend. Acc. Naz. Lincei. S. VIII vol. XIV **6**, 1—7 (1953). — [4] Limitazioni per gli invarianti di deformazione. Rend. Mat. e Appl. S. V, Vol. XIV, **3**, 269—279 (1955).

BOUSSINESQ, J.: [1] Note complementaire au mémoire sur les ondes liquides periodiques présenté le 29 novembre 1869, et approuvé par l'Académie le 21 fevrier 1870. Etablissement de relations générales et nouvelles entre l'énergie interne d'un corps fluide ou solide, et ses pressions ou forces élastiques. C.R. Acad. Sci. (Paris) **71**, 400—402 (1870). — [2] Théorie des ondes liquides periodiques. Mémoires présentés par divers savants à l'Académie des Sciences. **20**, 509—615 (1872).

BRESSAN, A.: [1] Sulle deformazioni dei corpi cristallini cilindrici nello schema di De Saint-Venant. Rend. Seminar. Mat. Univ. Padova, **22**, 281—293 (1953). — [2] Sull'integrabilità del problema di De Saint-Venant nei solidi cristallini. Rend. Seminar. Mat. Univ. Padova **23**, 435—448 (1954). — [3] Sulla possibilità di stabilire limitazioni inferiori per le componenti intrinseche del tensore degli sforzi in coordinate generali. Rend. Seminar. Mat. Univ. Padova **26**, 139—147 (1956).

BRILLOUIN, L.: Le tenseurs en Mécanique et en Élasticité. Masson et Cie Editeurs. Paris: 1938.

CAPRIOLI, L.: Su un criterio per l'esistenza dell'energia di deformazione. Boll. Un. Mat. Ital. S. III, **10**, 481—483 (1955).

CATTANEO, C.: Su un teorema fondamentale nella teoria delle onde di discontinuità. Rend. accad. nazl. Lincei. S. VIII, vol 1; N. 1, 66—72; N. 2, 728—734.

CHU, BOA-THE: Thermodynamics of elastic and of some visco-elastic solids and non-linear thermoelasticity. Tech. Rep. N. 1—562 (20) Division of Engineering, Brown University, Providence, R. I. pp. 44 (1957).

COLEMAN, B., and W. NOLL: On the thermostatics of continuous media. Arch. rational Mech. Anal. Vol. 4, N. 2, 97—128 (1959).

COLOMBO, G.: [1] Limitazioni superiori per i moduli delle componenti di stress in un particolare problema di deformazione piana. Ann. Un. Ferrara S. VII, V. III, **6**, (43—53) (1954). — [2] Maggiorazioni delle componenti di stress nel problema di De Saint-Venant. Rend. Seminar. Mat. Univ. Padova, **24**, 70—83 (1955).

CSONKA, P.: Contributions to the elastic theory of isotropic bodies. Acta Tech. Acad. Sci. Hungar **17**, 355—359 (1957).

DA SILVA, D. A.: Memoria sobre a rotaçao das forcas em torno dos pontos d'applicacao. Mem. Ac. R. Sci. Lisboa S. II, V. 3, p. I 61—231 (1851).

DE SAINT-VENANT: [1] Sur les pressions qui se developpent à l'interieur des corps solides lorsque les déplacements de leurs points, sans alterer l'élasticité, ne peuven tcependant pas être considérés comme très-petits. Bull. Soc. Philomath. 5, 26—28 (1844). — [2] Mémoire sur l'équilibre des corps solides, dans les limites de leur élasticité et sur les conditions de leur resistance quand les déplacements ne sont pas très-petits. C. r. Acad. Sci. (Paris) 24, 260—263 (1847).

DIAZ, J. B., and H. J. GREENBERG: Upper and lower bounds for the solution of the first boundary value problem of elasticity. Quart. Appl. Math. **6**, 326—331 (1948).

DUHEM, P.: Recherches sur l'élasticité. Ann. sci. école norm. super. 22, 203—206; **23**, 190—191 (1905).

ERICKSEN, J. L.: [1] Stress-deformation relations for solids. Canad. J. Physics **34**, 226—227 (1956). — [2] Hypo-elastic pure flexure. Quart. J. Mech. Appl. Math. **11**, 67—72 (1958).

FICHERA, G.: [1] Decomposizione al modo di Poincarè delle funzioni biiperarmoniche in due variabili. Rend. Accad. Sci. Fis. Mat. Napoli S. 4, n. 11 (1940—1941). — [2] Sull'equilibrio di un corpo elastico isotropo e omogeneo. Rend. Seminar Mat. Univ. Padova **27**, 9—28 (1948). — [3] Sui problemi analitici dell'elasticità piana. Rend. Fac. Sci. Cagliari **28**, 1—22 (1948). — [4] Sull'esistenza e sul calcolo delle soluzioni dei problemi al contorno, relativi all'equilibrio di un corpo elastico. Ann. Sc. Norm. Sup. Pisa S. III, V. IV, 1—65 (1950). — [5] Condizioni perchè sia compatibile il problema principale della Statica elastica. Rend. accad. nazl. Lincei, S. VIII, vol. XIV, **3**, 397—400 (1953). — [6] Sulla torsione elastica dei prismi cavi. Rend. Mat. e Appl. S. V, V. XII, 163—176 (1953).

FINGER, J.: Über die allgemeinsten Beziehungen zwischen Deformationen und den zugehörigen Spannungen in aeolotropen und isotropen Substanzen. Akad. Wiss. Wien Sitzungsber. (II a) **103**, 1073—1100 (1894).

FINZI, L.: Legame tra equilibrio e congruenza e suo significato fisico. Atti R. Accad. Lincei, (8), **20**, 205—211 (1956).

GASPARINI, I.: Sopra una proprietà caratteristica dei sistemi isotropi. Boll. Un. Mat. Ital. (2), I, 13—18 (1943).

GHIZZETTI, A.: Sugli stati di tensione piana in un corpo cilindrico elastico. Ann. Scuola Norm. Sup. Pisa, S. III, **5** (1951).

GREEN, A. E.: [1] Finite elastic deformation of compressible bodies. Proc. roy. Soc. A, **227**, 271—278 (1955). — [2] Hypo-elasticity and plasticity. Proc. roy. Soc. A, **234**, 46—59 (1956). — [3] Simple extension of a hypo-elastic body of grade zero. J. Rational Mech. **5**, 637—642 (1956).

GREEN, A. E., and E. B. SPRATT: Second order effects in the deformations of elastic bodies. Proc. roy. Soc. A, **224**, 347—361 (1954).

GRIOLI, G.: [1] Una proprietà di minimo nella Cinematica delle deformazioni finite. Boll. Un. Mat. Ital. S. II, A. II, **5**, 452—455 (1940). — [2] Struttura della funzione di Airy nei sistemi molteplicemente connessi. Giorn. Mat. Battaglini, 119—144 (1947). — [3] Sulle deformazioni elastiche di un involucro omogeneo soggetto a pressione o trazione. Rend. Seminar Mat. Univ. Padova, **20**, 278—285 (1951). — [4] Relazioni quantitative per lo stato tensionale di un qualunque sistema continuo e per la deformazione di un corpo elastico in equilibrio. Ann. Mat. pur. appl. Sez. IV, **33**, 239—246 (1952). — [5] Proprietà di media ed integrazione del problema dell'elastostatica isoterma. Ann. Mat. pur. appl. Ser. IV, **33**, 263—271 (1952). — [6] Validità del teorema di Menabrea e integrazione del problema dell'elastostatica in casi non isotermi. Rend. Seminar. Mat. Univ. Padova, **21**, 202—208 (1952). — [7] Sul problema di De Saint-Venant nei solidi cristallini. Rend. Seminar. Mat. Univ. Padova, **21**, 228—242 (1952). — [8] Integrazione del problema della Statica delle piastre omogenee di spessore qualunque. Ann. Scuola Norm. Sup. Pisa Ser. III, **6**, 31—49 (1952). — [9] Limitazioni per lo stato tensionale di un qualunque sistema continuo. Ann. Mat. Pura Appl. ser. IV, **39**, 255—266 (1955). — [10] Sullo stato tensionale dei continui in equilibrio e sulle deformazioni nel caso elastico. Conferenze Sem. Mat. Un. Bari 35—36 (1958). — [11] Elasticità asimmetrica. Ann. Mat. Pura Appl. Ser. IV, **50**, 389—417 (1960). — [12] Onde di discontinuità ed elasticità asimmetrica. Rend. Acc. Naz. Lincei. S. VIII, V. XXIX; f. 5 (1961).

HADAMARD, J.: Leçons sur la Propagation des Ondes et les Équations de l'Hydrodynamique. Paris: Hermann 1903.

HANDELMAN, LIN, and W. PRAGER: On the mechanical behavior of metals in the strain-hardening range. Quart. Appl. Math. **4**, 397—407 (1947).

HANIN, MIR, and M. REINER: On isotropic tensor-functions and the measure of deformation. Z. Angew. Math. Phys. **7**, 377—393 (1956).

HENCKY, H.: [1] Über die Form des Elastizitätsgesetzes bei ideal elastischen Stoffen. Z. tech. Phys. **9**, 214—223 (1928). — [2] Mathematical principles of rheology. Research **2**, 437—443 (1949).

JAUMANN, G.: Geschlossenes System physikalischer und chemischer Differentialgesetze. Akad. Wiss. Wien. Sitzungsber. (IIa) **120**, 385—530 (1911).

KIRCHHOFF, G.: Über die Gleichungen des Gleichgewichts eines elastischen Körpers bei nicht unendlich kleinen Verschiebungen seiner Theile. Akad. Wiss. Wien. Sitzungsber. **9**, 762—773 (1852).

KORN, A.: Solution générale du problème d'équilibre dans la théorie de l'élasticité dans le cas ou les efforts sont donnés à la surface. Toulouse Université (1908).

164 Bibliography

Kötter, F.: Über die Spannungen in einem ursprünglich geraden, durch Einzelkräfte in stark gekrümmter Gleichgewichtslage gehaltenen Stab. Sitzber. Preuss. Akad. Wiss. **2**, 895—922 (1910).

Lauricella, G.: Alcune applicazioni della teoria delle equazioni funzionali alla Fisica-Matematica. Nuovo Cimento **13**, 104, 155, 237, 501 (1907).
Lichtenstein, L.: Ueber die erste Randwertaufgabe der Elastizitäts-theorie. Math. Z. **20** (1924).
Locatelli, P.: Sul principio di Menabrea. Boll. Un. Mat. Ital. Vol. II Ser. II, 342—347 (1940).
Lodge, A. S.: [1] A new theorem in the classical theory of elasticity. Nature **169**, 926—927 (1952). — [2] The transformation to isotropic form of the equilibrium equations for a class of anisotropic elastic solids. Quart. J. Mech. Appl. Math. **8**, 211—225 (1955).
Love, A.: A Treatise on the Mathematical Theory of Elasticity. (1927).

Manacorda, T.: [1] Sul legame sforzi deformazione nelle trasformazioni finite di un mezzo continuo isotropo. Riv. Mat. Univ. Parma, **4**, 31—42 (1953). — [2] Sulla torsione di un cilindro circolare omogeneo e isotropo nella teoria delle deformazioni finite dei solidi elastici incomprimibili. Boll. Un. Mat. Ital. 177—189 (1955).
Michell, J. H.: Proc. London Math. Soc. Vol. 31, p. 100 (1899).
Milne-Thomson, L. M.: Plane Elastic Systems. Ergebnisse der angewandten Mathematik. Berlin, Göttingen, Heidelberg: Springer 1960.
Miranda, C.: Problemi d'esistenza in Analisi funzionale. Quaderni Mat. Scuola Norm. Sup. Pisa (1948—49).
Murnaghan, F. D.: [1] Finite deformations of an elastic solid. Am. J. Math. **59**, 235—260 (1937). — [2] The compressibility of solids under extreme pressures. Karman Ann. Vol. 121—136 (1941). — [3] The Foundations of the Theory of Elasticity. — Non-linear Problems in Mechanics of Continua N. Y. 158—174 (1947). — [4] A revision of elasticity. Anais Acad. Brasil **21**, 329—376 (1949).
Muskhelishvili, N. J.: Some Basic Problems of the Mathematical Theory of Elasticity. Noordhoff: Gronigen — Holland 1953.

Nadai: Elastische Platten.
Noll, W.: On the continuity of the solid and fluid states. J. Rat. Mech. Anal. **4**, 3—81 (1955).
Novozhilov, V. V.: [1] Foundations of the Nonlinear Theory of Elasticity. Graylock press. Rochester, N. Y. (1953). — [2] On the relation stress and strain in a nonlinear elastic medium. Akad. Nauk, SSSR. Prikl. Mat. Meh. **15**, 183—194 (1957).

Picone, M.: [1] Appunti di Analisi superiore. Rondinella, Napoli (1940). — [2] Nuovi metodi risolutivi per i problemi d'integrazione delle equazioni lineari a derivate parziali e nuova applicazione della trasformata multipla di Laplace nel caso delle equazioni a coefficienti costanti. Rend. Accad. Sci. Torino, 413—426 (1940). — [3] Lezioni di Analisi funzionale. Tumminelli, Roma (1946—47). — [4] Sulla traduzione in equazione integrale lineare di prima specie dei problemi al contorno concernenti i sistemi di equazioni lineari a derivate parziali. Rend. Accad. Naz. Lincei, Ser. VIII, **2**, (1947). — [5] Sulla torsione di un prisma elastico cavo secondo la teoria di Saint-Venant. Ann. Soc. Polonaise Math. **20**, 347—372 (1947). —

[6] Exposition d'une méthode d'intégration numerique des systèmes d'équations linéaires aux dérivées partielles mise en ouvre à l'Institut national pour les Applications du Calcul. Résultats obtenus et résultats que l'on pourrait attendre. Comptes rendus du Colloque International du Centre National de la Recherche Scientifique. «Les machines à calculer et la pensée humaine.» Paris 8—13 Janvier 239—261 (1951).

PIOLA, G.: La meccanica dei corpi naturalmente estesi trattata col calcolo delle variazioni. Opusc. mat. fis. di diversi autori. Milano: Giusti 1, 201—236 (1833).

PLATONE, M. G.: Sugli stati di tensione piana in un corpo cilindrico elastico. Ann. Scuola Norm. Sup. Pisa, Ser. III, 5, 57—70 (1951).

PLATRIER, C.: Conditions d'intégrabilité du tenseur de dèformation totale dans une transformation finie d'un milieu à trois dimensions. Acad. Roy. Belgique. Bull. Cl. Sci. (5) 39, 490—494 (1953).

PRAGER, W.: On the concept of stress rate. 10th International Congress of Applied Mechanics. Stresa, Italy (1960).

REISSNER, E.: Note on the theorem of the symmetry of the stress tensor. J. Math. Phys. 23, 192 (1944).

RIVLIN, R. S.: [1] Large elastic deformations of isotropic materials IV. Further developments of the general theory. Phil. Trans. R. Soc. (A) 241, 379—397 (1948). — [2] The solution of problems in second order elasticity. J. Rat. Mech. Anal. 2, 53—81 (1953).

RIVLIN, R. S., and J. L. ERICKSEN: Stress-deformation relations for isotropic materials. J. Rat. Mech. Anal. 4, 323—425 (1955).

RIVLIN, R. S., and D. W. SAUNDERS: Large elastic deformations of isotropic materials. VII. Experiments on the deformation of rubber. Phil. Trans. R. Soc. London (A) 243, 251—288 (1951).

SETH, B. R.: Finite strain in elastic problems. Phil. Trans. R. Soc. London (A) 234, 231—264 (1935).

SIGNORINI, A.: [1] Sopra alcune questioni di Statica dei sistemi continui. Ann. Scuola Norm. Sup. Pisa Ser. II, 2, 3—23 (1933). — [2] Trasformazioni termoelastiche finite. Atti XXIV Riunione della S.I.P.S. vol. 3, 17—18 (1935). — [3] Trasformazioni termoelastiche finite. Memoria Iª. Ann. Mat. pur. appl. Ser. IV, 22, 33—143 (1943). — [4] Trasformazioni termoelastiche finite. Memoria IIª. Ann. Mat. pur. appl. Ser. IV, 30, 1—72 (1949). — [5] Un semplice esempio di incompatibilità tra la Elastostatica classica e la teoria delle deformazioni elastiche finite. Rend. Acc. Naz. Lincei, Ser. VIII, 3, 276—281 (1950). — [6] Sopra un'estensione della teoria linearizzata dell'elasticità. Rend. Sem. Univ. Pol. Torino, 12, 83—93 (1952—53). — [7] Un'espressiva applicazione delle proprietà di media dello stress comuni a tutti i sistemi continui. Studies in mathematics and mechanics presented to Richard von Mises. Academ. Press Inc. N. Y. 274—277 (1954). — [8] Trasformazioni termoelastiche finite. Solidi incomprimibili. Memoria IIIª. Ann. Mat. pur. appl. Ser. IV, 39, 147—201 (1955). — [9] Questioni di elasticità non linearizzata e semilinearizzata. Rend. Mat. Roma 18, 1—45 (1959). — [10] Sulla Statica dei sistemi elastici vincolati. Atti VI Congresso dell'U.M.I. Napoli (1959). — [11] Questioni di elasticità non linearizzata. Rendiconti di Matematica (1—2) Vol. 19, p. 1—71 (1960). — [12] Trasformazioni termoelastiche finite. Solidi vincolati. Memoria IV. Ann. Mat. pur. appl. Serie IV, 51, 329—372 (1960).

SMITH, G. F., and R. S. RIVLIN: Stress-deformation relations for anisotropic solids. Arch. Rat. Mech. Anal. 1, 107—112 (1957).

SOBRERO, L.: [1] Theorie der ebenen Elastizität. Mathematische Einzel-
schriften. Hamburger 17, (1934). — [2] Nuovo metodo per lo studio dei
problemi di elasticità con applicazione al problema della piastra forata.
Ricerche di Ingegneria, A. II, 6, 1—12 (1934). — [3] Del significato
meccanico della funzione di Airy. Rend. Acc. Naz. Lincei 21, 264—269
(1935). — [4] Elasticitade. Livaria Boffoni. Rio de Janeiro (1942).

SOMIGLIANA, C.: Sopra un'estensione della teoria dell'elasticità. Rend. Acc.
Naz. Lincei 14, 43—50 (1910).

SOUTHWELL, R. V.: On the general theory of elastic stability. Phil. Trans.
R. Soc. London (A) 213, 187—244 (1913—1914).

STOPPELLI, F.: [1] Una generalizzazione di un teorema di Da Silva. Rend.
Acc. Sci. Fis. Mat. Soc. Naz. Sci. Lett. Arti Napoli Ser. 4, 21 (1954). —
[2] Un teorema di esistenza ed unicità relativo alle equazioni dell'elastostatica isoterma per deformazioni finite. Ricerche Mat. 3, 247—267
(1954). — [3] Sulla sviluppabilità in serie di potenze di un parametro
delle soluzioni delle equazioni dell'Elastostatica isoterma. Ricerche Mat.
4, 58—73 (1955). — [4] Su un sistema di equazioni integro-differenziali
interessanti l'Elastostatica. Ricerche Mat. 6, 11—26 (1957). — [5] Sull'-
esistenza di soluzioni delle equazioni dell'Elastostatica isoterma nel caso
di sollecitazioni dotate di assi di equilibrio. Memoria I. Ricerche Mat. 6,
241—287 (1957). — [6] Sull'esistenza di soluzioni delle equazioni dell'Elastostatica isoterma nel caso di sollecitazioni dotate di assi di equilibrio.
Memoria II, 7, 71—101 (1958). — [7] Sull'esistenza di soluzioni delle
equazioni dell'Elastostatica isoterma nel caso di sollecitazioni dotate di
assi di equilibrio. Memoria III. 7, 138—152 (1958).

SUPINO, G.: Sopra alcune limitazioni per la sollecitazione elastica e sopra la
dimostrazione del principio di De Saint-Venant. Ann. Mat. pur. appl.
Ser. IV, 9, 91—119 (1931).

SYNGE, J. L.: Upper and lower bounds for the solutions of problems in
elasticity. Proc. Roy. Irish Acad. Sect. A. 53, 41—64 (1950).

THOMAS, T. Y.: Deformation energy and the stress-strain relations for isotropic materials. J. Math. Phys. 35, 335—350 (1957).

THOMSON, W. (Lord KELVIN): Dynamical problems regarding elastic
spheroidal shells and spheroids of incompressible liquid. Phil. Trans. R.
Soc. London (A) 153, 583—616, Papers 3, 351—394 (1863).

TIMOSHENKO, S., and T. GOODIER: Theory of Elasticity. Engineering So-
cieties Monographs Mc Graw-Hill Book Company, Inc. Second edition,
(1951).

TIMPE, A.: Probleme der Spannungsverteilung in ebenen Systemen, einfach
gelöst mit Hilfe der Airyschen Funktion. Z. Math. Phys. 52, p. 348 (1905).

TOLOTTI, C.: [1] Sui problemi di elasticità piana a funzione di Airy poli-
droma. Rend. Acc. Naz. Lincei. Serie VI, Vol. XXV, 226—230 (1937). —
[2] Orientamenti principali di un corpo elastico rispetto alla sua sollecitazione totale. Mem. Acc. Ital. Ser. VII, 13, 1139—1162 (1942). —[3] Sulla
più generale elasticità di 2° grado. Rend. Mat. Appl. Ser. V, 3, 1—20
(1942). — [4] Sul potenziale termodinamico dei solidi elastici omogenei
ed isotropi per trasformazioni finite. Atti Acc. Ital. 14, 529—541 (1943). —
[5] Deformazioni elastiche finite: onde ordinarie di discontinuità e casi
tipici di solidi elastici isotropi. Rend. Mat. Appl. Ser. V, 4, 34—59 (1943).

TONOLO, A.: [1] Forma intrinseca delle equazioni dell'equilibrio dei mezzi elastici. Rend. Acc. Lincei **11**, 247—250; 347—351; 652—654 (1930). — [2] Teoria tensoriale delle deformazioni finite dei corpi solidi. Rend. Sem. Mat. Padova, **14**, 1—75 (1943).

TRUESDELL, C.: [1] A new definition of a fluid. J. math. pures appl. (9), **30**, 111—158 (1951). — [2] The mechanical foundations of elasticity and fluid dynamics. J. Rat. Mech. An. **1**, 125—300 (1952). — [3] Corrections and additions to the Mechanical foundations of elasticity and fluid dynamics. J. Rat. Mech. An. **2**, 593—616 (1953). — [4] The simplest rate theory of pure elasticity. Comm. Pure Appl. Math. **8**, 123—132 (1955). — [5] Hypo-elasticity. J. Rat. Mech. Anal. **4**, 83—133 (1955). — [6] Hypo-elastic shear. J. Appl. Phys. **27**, 441—447 (1956). — [7] L'ipo-elasticità. Conferenze Sem. Mat. Univ. Bari **29** (1957). — [8] The rational mechanics of materials — past, present, future. Appl. Mech. Rev. **12**, 75—80 (1959). — [9] General and exact theory of waves in finite elastic strain. Arch. Rat. Mech. Anal. **8**, 263—296 (1961).

UDESCHINI, P.: Sull'energia di deformazione. Rend. Ist. Lombardo di Sci. Lett. Ser. III, **76**, 25—34 (1943).

VERMA, P. D. S.: [1] Hypo-elastic pure flexure. Proc. Indian Acad. Sci. Sect. A. **44**, 185—193 (1956). — [2] Hypo-elastic strain in a rotating shaft and a spherical shell. Proc. Second Congress Theor. Appl. Mech. New Delhi 99—110 (1956). — [3] Deformation energy for hypo-elastic materials of grade zero. J. Sci. Eng. Res. India **2**, 251—252 (1958).

VOIGT, W.: [1] Theoretische Studien über die Elasticitätsverhältnisse der Krystalle. Abhnd K. Ges. Göttingen (1887). — [2] Ueber eine anscheinend nothwendige Erweiterung der Elasticitätstheorie. Gott. Nach. 534—552 Ann. der Phys. (2) **52**, 536—555 (1893). — [3] Ueber eine anscheinend nothwendige Erweiterung der Elasticitätstheorie. Gott. Abh. 33—43 (1895).

VOLTERRA, V.: Sur l'équilibre des corps élastiques multiplement connesses. Gauthier-Villars (1907).

VON MISES, R.: Mechanik der festen Körper im plastisch-deformablen Zustand. Nachr. Ges. Wiss. Göttingen (1913).

ZAREMBA, S.: Sur une forme perfectionée de la théorie de la relaxation. Bull. Int. Acad. Sci. Cracovie 594—614 (1903).

ZVOLINSKI, N. V., and P. M. RIZ: O nekotorykh zadachakh neilineinoi teorii uprugosti. Prikl. Mat. Mekh. (2) **2**, 417—426 (1938—1939).

Index of Authors